THE CD-ROM DRIVE
A Brief System Description

THE CD-ROM DRIVE
A Brief System Description

SORIN G. STAN

Philips Optical Storage
Optical Recording Development Laboratory
Eindhoven, The Netherlands

Kluwer Academic Publishers
Boston/Dordrecht/London

A C.I.P. Catalogue record for this book is available from the Library of Congress.

ISBN 978-1-4419-5039-0

Published by Kluwer Academic Publishers,
P.O. Box 17, 3300 AA Dordrecht, The Netherlands.

Sold and distributed in North, Central and South America
by Kluwer Academic Publishers,
101 Philip Drive, Norwell, MA 02061, U.S.A.

In all other countries, sold and distributed
by Kluwer Academic Publishers,
P.O. Box 322, 3300 AH Dordrecht, The Netherlands.

Printed on acid-free paper

Contents

List of Figures

Preface

The Compact Disc (CD), as a standardized information carrier, has become one of the most successful consumer products ever marketed. Although the original disc was intended for audio playback, its specific advantages opened very quickly the way towards various computer applications. The standardization of the Compact Disc Read-Only Memory (CD-ROM) and of all succeeding similar products, like Compact Disc interactive (CD-i), Photo and Video CD, CD Recordable (CD-R), and CD Rewritable (CD-R/W), has substantially enlarged the range of possible applications.

The plastic disc represented from the very beginning a removable medium of large storage capacity. The advent of the personal computer accompanied by the increasing demand for both data distribution and exchange have strongly marked the evolution of the CD-ROM drive. The number of sold CD-ROM units exceeded 60 millions in 1997 when compared to about 2.5 millions in 1992.

As computing power continuously improved over the years, computer peripherals have also targeted better performance specifications. In particular, the speed of CD-ROM drives increased from the so-called 1X in 1984 to double speed in 1992, and further to 32X at the beginning of 1998. The average time needed to access data on disc has dropped from about 300 ms to less than 90 ms within the same period of time.

This book tries to provide a comprehensive description of the CD-ROM drive as a complex system. Without emphasis on mathematical and physical details, the text provides the backgrounds necessary to understand the functionality of the drive. The aspects concerning current high-speed drives and the related design issues are also discussed. When extrapolated to other optical storage systems, like CD-R, CD-R/W, and even the emerging Digital Versatile Disc (DVD), the book should represent a valuable reference. A large bibliography list is attached at the end of the book and the numerous references to this list which are made throughout the text may help the reader finding the desired details. The book was written with the clear goal in mind of being equally useful to specialists in the field and readers with little or no CD-ROM experience.

The first chapter of the book introduces the reader to the world of disk-based mass storage devices and tries to position the CD-ROM within this framework. A brief history of Compact Disc Digital Audio, an overview of the CD standards and the advantages of compact disc are presented. The chapter ends with a short system description of the CD-ROM drive.

The following three chapters are dedicated to the main subsystems that cooperate to provide the binary read-out signal. These chapters describe, in order, the optics, the servomechanics, and the decoder circuitry.

The fifth chapter discusses those building blocks of the drive which are indispensable to the disc read-out and either link the main three subsystems mentioned above or represent a sort of terminator in the whole system. We approach here the laser diode, the photodetector and preprocessing circuits, the audio part, and the drive microcontroller.

The sixth chapter provides information about the signal processing unit which is specifically dedicated to both deal with CD-ROM data and interface the drive with the host PC.

Finally, the last chapter presents an overview of several system parameters and of the design choices made more than 15 years ago when the CD standards were set. This chapter ends with some issues related to the performance specifications and benchmarking of CD-ROM drives.

This book could not have arrived at its final stage without the contribution of several people. First, I would like to acknowledge the outstanding professional and moral support I had from Prof. Paul van den Bosch and Assoc. Prof. Ad Damen from the Eindhoven University of Technology, and from Dr. Maarten Steinbuch from Philips Research Laboratories in Eindhoven. They encouraged my efforts towards a more complete work related to the CD-ROM system from which this book has actually emerged as a spin-off. They also read the whole manuscript of the book, suggested improvements and contributed to its consistent structure.

Special thanks go to Hans Naus from Optical Recording Development Laboratory of Philips Optical Storage who understood the goal of my work, believed in its importance for myself and for our company, and supported it throughout the Philips organization. I am also indebted to the Philips management which allowed me to publish this book.

I am extremely grateful to Dr. Stan Baggen, Dr. Joseph Braat and Dr. Kees Schouhamer-Immink from Philips Research Laboratories in Eindhoven who scrutinized parts of the manuscript and suggested valuable improvements of its contents. It is also a pleasure to acknowledge the unexpected meeting with Prof. J. Peek from University of Nijmegen who reacted very enthusiastically to my work, including this book as a spin-off.

Many thanks go to my office colleague Ton Akkermans with whom I shared many of my thoughts and discussed plenty of technical aspects related to optical recordings. He read and commented parts of the manuscript.

I am also indebted to my colleagues Peter Coops, Conor O'Gorman, Cees Hezemans, Henk van Kempen, Aart Lighart, Job van Mil, Han Raaijmakers, and Cees Smulders from Optical Recording Development Laboratory of Philips Optical Storage in Eindhoven. They provided inspiration and improved the quality of the manuscript. I would also like to thank George Leenknegt and Peter de la Rambelje from the same laboratory and Eric van Rheden from the Business Intelligence office of Philips Optical Storage for their moral encouragement in carrying out my work. Thanks go also to Hilde Overath and Rob Groen, former managers I had, who gave me the opportunity to work on CD-ROM drive development.

I would like to express my gratitude to Prof. Yossi Chait from University of Massachusetts, Dr. Yuan Zheng from Oak Technology, Inc., and James Finlay from Kluwer Academic Publishers for their positive comments on the need for a CD-ROM monograph.

This book is the result of more than one year of efforts made independently from my current tasks within the Philips company. During our weekends spent together, my son Robert provided me with the necessary stress relaxation. I really enjoyed the evening walks together, between my daily activities at Philips and the work I carried out at home. Last but not least, I am indebted to my wife for her patience, understanding, dedication, and willingness to sacrifice many evenings in favor of my professional results.

Sorin G. Stan
Eindhoven, The Netherlands
March 1998

P.S. Several minor errors and a mistake in Equation (4.5) have been corrected in this reprinted version of the book.

Glossary

Acronyms and Abbreviations

ADC	Analog-to-Digital Conversion (or Converter)
AlGaAs	Aluminum-Gallium-Arsenide
BCLK	Bit CLocK
BER	Bit Error Rate
BERL	Burst ERror Length
BLER	BLock Error Rate
CA	Central Aperture
CAV	Constant Angular Velocity
CDBD	Compact Disc Block Decoder
CD-DA	Compact Disc Digital Audio
CD-i	Compact Disc interactive
CDM	Compact Disc Mechanism
CD-R	Compact Disc Recordable
CD-ROM	Compact Disc Read-Only Memory
CD-ROM/XA	CD-ROM eXtended Architecture
CD-RW	Compact Disc ReWritable
CD-R/RW	Compact Disc Recordable/ReWritable
CIRC	Cross-Interleaved Reed-Solomon Code
CLV	Constant Linear Velocity
CPU	Central Processing Unit
CRC	Cyclic Redundancy Check
CW	Continuous Wave
D/A	Digital-to-Analog
DAC	Digital-to-Analog Conversion (or Converter)
DAT	Digital Audio Tape
DC	Direct Current
DDS	Digital Data Storage

DH	Double-Heterostructure (laser)
DMA	Direct Memory Access
DSP	Digital Signal Processing (or Processor)
DTO	Digitally-Controlled Oscillator
DVD	Digital Versatile Disc
ECC	Error Correction Code
EDC	Error Detection Code
EFM	Eight-to-Fourteen Modulation
emf	electromotive force
FE	Focus Error
FIFO	First-In-First-Out
FWHM	Full Width at Half Maximum
GaAs	Gallium-Arsenide
HF	High Frequency
HOE	Holographic Optical Element
I^2S	Inter-IC Sound
IC	Integrated Circuit
IDE	Intelligent Drive Electronics
ISI	InterSymbol Interference
laser	light amplification by stimulated emission of radiation
LDGU	Laser Diode Grating Unit
LPF	Low-Pass Filter
LSB	Least Significant Bit (or Byte)
MB	MegaByte
MO	Magneto-Optical
MSB	Most Significant Bit (or Byte)
MTF	Modulation Transfer Function
NA	Numerical Aperture
NCU	Nonlinear Control Unit
NRZ	NonReturn-to-Zero
OPU	Optical Pickup Unit
PC	Personal Computer
PCB	Printed Circuit Board
PD	Phase-change Dual
PDM	Pulse Density Modulation

PID	Proportional, Integral and Derivative
pin	p-type, intrinsic, n-type semiconductor device
PLL	Phase-Locked Loop
pn	p-type, n-type junction (or semiconductor device)
QFT	Quantitative Feedback Theory
RE	Radial Error
RLL	Run-Length Limited
RMS	Root Mean-Squared
rpm	rotations per minute
RS	Reed-Solomon (code)
SCSI	Small Computer System Interface
SI	Système International d'Unités
SNR	Signal-to-Noise Ratio
SSD	Spot-Size Detection
VCO	Voltage-Controlled Oscillator
WCLK	Word CLocK
XOR	eXclusive OR
Xtal	Crystal clock
WORM	Write-Once-Read-Many
$\Sigma\Delta$	Sigma-Delta (Modulation or Conversion)

Symbols[1]

A_0	optical amplitude of the zeroth diffracted order
A_1	optical amplitude of the first diffracted order
A_D	detection level for the HF signal, in V
B_a	user bit rate of a CD-ROM disc rotating at 1X CLV, approximately equal to 150 kB/s
$\overline{B}_{av}$	average bit rate, in kB/s
C_{cdrom}	CD-ROM storage capacity, in MB
d_{al}	thickness of the active layer in a laser diode, in μm
d_m	minimum distance of a given code
D	damping (viscous friction) constant, in Ns/m for linear displacement or Nm/s^{-1} for rotational movement

[1]Where appropriate, either the SI unit or an SI derivative which is commonly used is given.

D_i	inner diameter of the program area, in mm
D_o	outer diameter of the program area, in mm
D_p	diffusion coefficient for holes in the n-region of a photodiode, in cm^2/s
f	frequency, in Hz
f_c	cross-over frequency of an open-loop amplitude characteristic, in Hz
f_{ch}	channel bit rate, in Mbit/s
f_{kT}	frequency of a kT EFM pattern, in kHz
f_n	natural undamped frequency of a second-order system, in Hz
f_{rot}	disc rotational frequency, in Hz
F_s	sampling frequency of digital audio, equal to 44.1 kHz
F_{unbal}	unbalance force generated by the rotating disc, in mN
g_0	gain constant defining the optical gain per unit length of a laser diode, in cm^{-1}
$\hbar$	Plank constant, equal to $6.6256 \cdot 10^{-34}$ Js
I_3	peak-peak amplitude of the HF component having the highest fundamental frequency, in A or V
I_{11}	peak-peak amplitude of the HF component having the lowest fundamental frequency, in A or V
I_{pp}	maximum variation of photodetector current, in A
I_{top}	peak value of the HF signal before high-pass filtering, in A or V
J_0	nominal current density defining the optical gain of a semiconductor laser, in A/cm^2
J_{rot}	rotational moment of inertia at the DC motor shaft, in kgm^2
J_t	threshold current density in a laser diode, in A/cm^2
K	elastic constant (stiffness) of a spring, in N/m
K_e	back-emf constant, in Vs/m for linear displacement or Vs for rotational movement
K_d	derivative gain of a PID controller
K_f	force constant (of a voice-coil motor), in N/A
K_i	integral gain of a PID controller
K_p	proportional gain of a PID controller
K_t	torque constant (of a rotary motor), in Nm/A
L_{al}	length of the active layer of a semiconductor laser, in μm

L	inductance (of a coil), in H
L_a	armature inductance (of a motor coil), in H
L_{bit}	physical length of the channel bit, in nm
L_{kT}	physical length of a pit/land, measured along the disc spiral and corresponding to a kT EFM pattern, in nm
m	order of the diffracted beam
M	radial modulation index
M_{act}	moving mass of the actuator, in kg
n_i	refraction indices of the semiconductor layers ($i = 1, 2, 3$) of a DH laser diode
n_{sub}	refractive index of the transparent substrate
N_1, N_2	oversampling factors for $F_s = 44.1$ kHz of the digital audio
N_b	number of data bits stored on an optical disk
$\mathcal{N}$	overspeed (X-factor) of the disc scanning velocity with respect to the audio reference velocity v_a
NA	numerical aperture (see also the list of abbreviations)
p	spatial period of the disc pit/land structure, in lines/μm or μm^{-1}
P_i	parity symbols for C_1 decoder, $i = 1, 2, 3, 4$
q	track pitch, in μm
Q	quality factor of a second-order system
Q_i	parity symbols for C_2 decoder, $i = 1, 2, 3, 4$
r	disc radius at the current read-out point, in mm
R	electrical resistance, in Ω
R_a	armature resistance (of a motor coil), in Ω
R_{coat}	reflectance of the coating surface of a photodiode
R_{mirr}	reflectance of the mirrors of a laser diode
R_{Airy}	radius of the Airy disc, in μm
s	Laplace (complex) variable
S_{final}	final (target) subcode timing after a seek action, in s
S_{init}	initial subcode timing before a seek action, in s
S_{tot}	total subcode timing on disc (maximum playback time at 1X CLV), in s
S_x, S_y	subcode timing, in s
t	time, in s
t_r	rise time of the photodiode impulse response, in s

T_{access}	access time, in s
T_{lat}	latency time, in s
$\overline{T}_{lat}$	average latency time, in s
T_{mot}	turntable motor time, in s
T_{ov}	overhead time, in s
T_r	retry time, in s
T_{seek}	seek time, in s
v_a	linear velocity of recorded data, in m/s
W	width of the photodiode depletion layer, in μm
x	linear displacement, in m
α	loss in optical energy (absorption coefficient) per unit length within a laser diode or photodiode, in m^{-1}
ΔE	difference between energy levels in a photonic device, in eV
ΔN_{tr}	number of tracks crossed during a seek action
$\overline{\Delta N}_{tr}$	average number of tracks crossed during a seek action
ΔR_{gr}	distance between the disc center of gravity and the rotation axis, in mm
$\overline{\Delta S}$	average seek length, in s
ΔS_{foc}	low-frequency sensitivity of the focus loop
ΔS_{rad}	low-frequency sensitivity of the radial loop
Δz	focal depth, in μm
Δx	spatial frequency for the MTF, in lines/μm or μm^{-1}
η	quantum efficiency of a photonic emitting device
η_{rec}	recording efficiency
η_{ext}	external quantum efficiency of a photodetector
η_{in}	internal quantum efficiency of a photodetector
θ_0	angle between the direction of the zeroth order and the direction normal to the incidence plane, in rad
θ_2	reflection angle inside the resonator of a laser diode, in rad
θ_m	angle between the direction of the mth order and the direction normal to the incidence plane, in rad
λ	wavelength of the laser light, in nm
ξ	damping ratio of a second-order system
τ	time constant, in s
τ_{el}	electrical time constant, in s
τ_p	lifetime of excess holes in the n-region of a photodiode, in s

φ_m	phase margin of a control loop, in rad
Γ	confinement factor within a semiconductor laser
Φ_{spot}	diameter of the laser spot, in μm
ψ_{10}	phase shift between zeroth and first diffracted orders, in rad
ω	angular frequency, in rad/s
ω_d	disc angular frequency, in rad/s

Notational conventions

$(2,10)$	RLL binary sequence with at least 2 and at most 10 zeros between any two consecutive ones (EFM code)
$3T$	smallest EFM pattern (see also kT below)
$B(S_x)$	user bit rate measured at the subcode timing S_x
C_1	first error detection and correction code (CIRC)
C_2	second error detection and correction code (CIRC)
$d(s)$	internal/external disturbances affecting a control loop
$e(s)$	error compensated by a closed control loop
$e_t(s)$	tracking error in a closed control loop
$F(t)$	force, as a function of time
$g(x)$	generator polynomial of a CRC code
$G_1(s)$	transfer function of the error detector in a control loop
$G_2(s)$	transfer function of the power driver in a control loop
$GF(q)$	Galois field with q elements
$H(s)$	open-loop transfer function, in general
$H_{act}(s)$	actuator transfer function
$H_{mot}(s)$	transfer function of the turntable motor
$i(t)$	electrical current, as a function of time
j	$\sqrt{-1}$
$I_{det}(t)$	intensity of the light received by the CA photodetector, as a function of time
$K(s)$	transfer function of the compensation network
kT	EFM pattern of $k-1$ zeros between two ones
$MTF(\Delta x)$	modulation transfer function, as dependent on the spatial frequency
$n(s)$	measurement noise affecting a control loop

(n, k)	RS code of length n and dimension k
$\mathcal{N}(S_x)$	overspeed factor measured at the subcode timing S_x
$\mathrm{RE}(x)$	radial error signal, as function of the radial displacement x
$S(s)$	sensitivity of a closed control loop
$\mathrm{SAT}_k(x, x_0)$	signal delivered by one of the lateral photo-detectors $(k = 1, 2)$
$(t = 2, e = 4)$	CIRC error detection and correction system with at most 2 direct corrections at C_1 level and at most 4 erasure corrections at C_2 level
$u_a(t)$	applied voltage, as a function of time
$u_b(t)$	back emf, as a function of time
$U_a(s)$	applied voltage, in Laplace domain
$U_b(s)$	back emf, in Laplace domain
$x(t)$	linear displacement, as function of time
$X_a(s)$	actuator position with respect to its carriage (sledge)
$X_d(s)$	position of a read-out point on the disc surface
$X_s(s)$	spot position as controlled by a focus/radial servo loop

Chapter 1

The CD-ROM Challenge

1.1 Introduction

The necessity of a storage device has always been of paramount importance for any computing system. In this respect, two categories of information need to be stored on some dedicated information carrier.

First, the executable programs must be preserved between separate computing sessions, when the power would most probably be switched off. Second, there are usually relatively large amounts of data which are used as input for the executable programs and are generated by the computing process itself.

The storage devices have basically followed two directions of development, determined by the speed at which data can be recorded and retrieved. Accordingly, these devices can be quite fast (e.g. semiconductor memories) or relatively slow, like the disk-based and the magnetic tape drives. The latter are referred to as mass storage devices since they allow easily for transport and exchange of huge quantities of data. Clearly, the improvement of the data recording and/or reading speed has always been set as one of the main performance specifications and has pushed ahead the development of such devices. On the other hand, the amount of data that can be stored by these devices represents another common specification point, irrespective of their read/write speed. From this perspective, the evolution of all storage systems has been marked by a continuously increasing storage capacity.

Other issues that have played and still play a very important role during the development of the storage systems are the removability of the information carrier itself (e.g. the disk), the eventual standardization, the protection of the storage medium against handling and climatic variations, etc.

The disk-based storage systems form a category of storage devices to which the Compact Disc[1] Read-Only Memory (CD-ROM) drive also belongs. When compared to, for instance, the hard-disk, the CD-ROM as a storage medium is exclusively intended for reading the recorded information and not for recording purposes[2]. However, the CD-ROM and the related drive have experienced an enormous success within a very short period of time. Partly, this success is due to the advent of the personal computer (PC) and partly due to the continuous and fast changes taking place in information technology [37]. The advantages offered by the CD-ROM storage have also contributed, clearly, to its success.

1.2 High-capacity storage devices

First of all, it is important to notice that the attribute of high capacity as well as the definition of a mass storage device are both of a relative nature. The PC hard-disk from 15 years ago could only record 20 MB of data and was still considered, in those times, a mass storage drive. This history will repeat itself again and again but perhaps, one of the most appropriate examples concerning the subject of this book is the introduction of the Digital Versatile Disc (DVD) and the decay of the "high-capacity" CD-ROM.

Before considering the CD-ROM technology, it is interesting to briefly explore the large variety of storage media nowadays on the market. The

---

[1]In this particular context, the UK English term "disc" has been chosen as an alternative for the American "disk", mainly to comply with the related standards [50,53,79,80]. The latter term will however be used throughout this book to designate other storage devices and media (e.g. hard-disk) than those based on a compact disc.

[2]Current standards are also defining a CD-Recordable (CD-R) and a CD-Rewritable (CD-RW) media which, after being written, feature both the same data structure as a CD-ROM and can be read with a CD-ROM drive.

reason for this approach is given by the common driving forces that push ahead the development of the disk-based storage devices. A comparative overview of several disk systems is presented in Table 1.1. The performance indicators (i.e., sustained transfer rate and average access time) used in this table will be explained in Section 7.4.

Storage device	Storage capacity [MB]	Sustained transfer rate [kB/s]	Average access time [ms]
Reference 1X CD-ROM drive	650	153.6	250 - 300
Standard floppy-disk	1.44	250/500	90
LS-120 floppy-disk	120	400 - 680	70
PD drive	650	300 - 1200	100 - 200
Iomega Zip drive	100	800 - 1400	30
Reference DVD drive	4700/8500 [1]	1385	150 - 200
DDS-3 DAT magnetic tape	12000 [2]	2000 - 2500	–
Magneto-optical drive	128 - 2600	1000 - 4500	20 - 40
Current CD-ROM drives	650	2000 - 4800	85 - 125
Iomega Jaz drive	2000	3400 - 6600	17
SyJet removable hard-disk	1500	3700 - 6900	12
Fixed hard-disk	1000 - 47000	8000 - 40000	8 - 15

[1] Single and double layer, respectively.
[2] Native mode (uncompressed).

Table 1.1   Basic performance indicators for the most common high-capacity storage devices, listed according to their sustained transfer rate.

As already mentioned at the beginning of this chapter, there has always been a clear trend towards increasing the amount of data which can be stored at once by a given device. This development direction goes not only hand in hand with the size of the available software but also with the amount of data needed to be processed by a given application. In addition, as the PC computing power has really sky-rocketed within the last few years, the PC-based applications have also become more demanding, giving rise to the necessity of storing, for instance, huge quantities of experimental data.

The speed at which data can be retrieved is also considered in Table 1.1. Although many devices have also a record function, their read performance is taken into account in order to facilitate the comparison with the read-only drives. At this stage, a difference should be made between continuous read-out and access of the desired recorded data.

The read-out speed, commonly designated as sustained data rate or data throughput, represents a measure of the amount of data transferred towards a host system[3] within a given time interval. This requirement is usually expressed in kilobytes (kB) or Megabytes[4] (MB) per second. Within the last years, the data throughput of the current storage systems has been pushed in the neighborhood of several MB/s.

Last but not least, the speed at which the recorded information can be accessed is crucial for computer applications. Hand in hand with the increasing processing power of the processors (also designated as central processing units or CPUs), the access time to the available information has also dropped. The faster the processor is, the quicker it needs data to handle. For small amounts of data, the speed of the semiconductor memories has closely followed the increasing speed of the microprocessors. Although for large amounts of data it is quite difficult, if not impossible, to replicate this speed, we are enjoying now access times of about 8 milliseconds for a typical hard-disk.

1.3 CD and CD-ROM history

The storage devices from Table 1.1 are ranked according to their sustained transfer rate. This choice is dictated by the evolution of the CD-ROM drives. Ever since their introduction, the so-called X-factor has determined this performance indicator and has perpetually driven the development of these drives.

1.3.1 The Compact Disc Digital Audio (CD-DA)

The CD system represents a complex conglomerate of technologies pioneered by many individuals and companies [83]. The primary development of the system itself is, however, attached to the names of two giants of electronics: Philips and Sony. The latter company is credited with the development of the error correction technique (and manufactured the very first integrated circuit dedicated to CD error correction) while Philips is considered the inventor of the optical technology employed in the compact disc.

In 1972 Philips announced the technique of storing audio recordings on an optical disc. A small diameter for the disc was fixed as a design requirement

---

[3]A disk storage device is considered as a peripheral of a central host system.

[4]1 MB = 1024 kB = 1024^2 bytes.

and analog modulation methods were being used. Sony was also exploring the possibilities of optically recording audio information on a larger disc and was looking for practical error-correction possibilities.

By 1977, many other companies had already shown prototypes of audio players based on an optical disc. In 1978, Philips and Sony agreed upon the signal format and error-correction methods and one year later they decided to further extend their collaboration with respect to other compact disc issues. They jointly proposed in June 1980 the Compact Disc Digital Audio system which was adopted in 1981, as a standard, by the Digital Audio Disc Committee, representing more than 25 manufacturers.

The introduction of the very first audio player was delayed for the next two years because dedicated signal processing integrated circuits (ICs) as well as semiconductor laser pickup units were not available in mass quantities. The introduction of the compact disc system took place simultaneously in Japan and Europe, in October 1982.

1.3.2 Compact Disc standards

Since the CD-DA standard [50,79] was established in 1981, a couple of other spin-off standards have also been set. They are currently referred to as Red Book, Yellow Book, etc. and cover specific CD-based applications or extend a previously established standard. Significantly, not all standards were proposed by Philips and Sony (the CD Video or Laser Vision, for instance, was defined by Pioneer while Photo CD was announced by Eastman Kodak). Other standards were jointly developed by more companies, such as Matsushita, Philips, Sony and JVC in the case of Video CD, or Philips, Microsoft and Sony in the case of CD-ROM/XA[5]

A simple diagram of several CD standards is depicted in Fig. 1.1. Additional details can be found in [85] or [54,83,116]. However, some authors only refer to the year when the product itself was launched on the market. For CD-DA and CD-ROM, for instance, the market introduction took place later on, after standardization.

1.3.3 The growth of the CD-ROM

Once the standard was established, the CD-ROM drives began to penetrate the computer markets. In the early years of the personal computer, when

---

[5]CD-ROM Extended Architecture

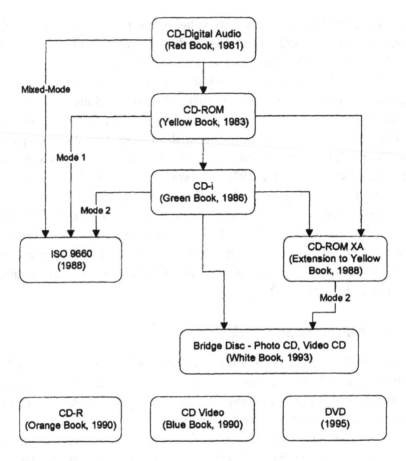

Figure 1.1   An overview of the optical disc standards and their related standardization years.

640 kB of PC internal memory was "enough for everyone", a CD-ROM storage capacity of 650 MB would have seemed rather exaggerated.

However, from a PC add-on which was separately sold in 1990, the CD-ROM drive has established itself as a standard PC component before the end of this millennium. Software houses sell their products, not anymore on floppy disks, but on CDs. Huge archives are regularly transferred onto compact discs and, last but not least, the entertainment world (games, video applications, etc.) has also boosted their business due to this optical medium.

The sales of CD-ROM discs have followed a spectacular growth and so did the sales of CD-ROM drives too. The first CD-ROM drives came into the market in 1985 but they really took off in the early nineties and reached volume sales several years later. An overview of estimated world-wide sales [22,28,33,75] since 1988 is presented in Fig. 1.2.

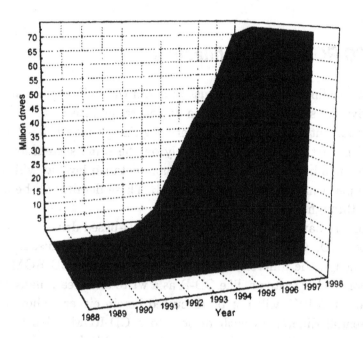

Figure 1.2   Estimated world-wide sales of CD-ROM drives.

1.3.4   The advantages of CD storage

The compact disc, as an optical medium, has definitely brought several specific advantages into the data storage arena.

First of all, the plastic disc has offered from the very beginning a support for storing relatively large amounts of information. As a medium, the disc itself was small, light and could easily be handled. Along with its high storage capacity, these characteristics opened a new way for porting data and applications between computers and between their users. In this respect, one of the main strengths of the compact disc was its accepted standardization, which can be translated into world-wide compatibility. Second, the recorded data is not affected by dust and normal fingerprints, making the CD extremely suitable for software distribution or any other type of data exchange.

There is also no disc degradation taking place during playback, no matter how often the disc is being used. And last but not least, the quality of the recorded information remains unchanged in time, even under large climatic variations.

1.4   CD-ROM drive architecture

Before proceeding further to discuss in detail the building blocks of a CD-ROM drive, we shall briefly describe its general architecture.

It is important to notice from the very beginning that all CD-ROM drives, independent of their manufacturer, rely on the same system architecture. The differences between drives lie at a more detailed level at which, in fact, the drive performance will be determined. These details will be discussed, however, throughout the next chapters.

A second remark concerns other types of CD-based drives, such as Video and Photo CD, CD-i, etc. These systems will not be discussed in this book, although the differences between them and the CD-ROM drive are very minor. Other systems, like CD-R as a write-once-read-many (WORM) system and CD-RW, will not be discussed herein either. They do feature more essential differences with respect to a CD-ROM drive but are still based on many similar, if not the same, building blocks.

A block diagram of the CD-ROM drive, as a system, is presented in Fig. 1.3. In general, the drive can be divided into a basic engine and a data path. These two generic parts are symbolically separated in Fig. 1.3 by a bus of control signals. Apart from other specific CD-ROM functions, the data path provides also the interface between the basic engine and the host system (usually a PC).

The rotating disc is read out without any mechanical contact with its surface. An optical pickup unit (OPU) sends a laser beam to the disc and receives back a reflected beam which has been optically modulated by the disc geometrical structure. The optical pickup unit contains, among other components, a semiconductor laser, optical elements to guide the laser beam, and a photodetector to transform the optical power into photocurrents. By properly processing the photocurrents, two servo signals can be derived for positioning the laser beam along the disc radius and spiral, respectively. At the same time, a high-frequency signal carrying the information recorded on the disc is also extracted and forwarded to the decoding electronics.

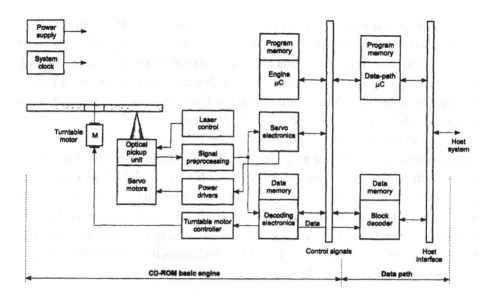

Figure 1.3 Simplified block diagram of the CD-ROM architecture.

The displacement of the laser beam along the vertical and radial directions with respect to the disc is accomplished by two tiny motors. They are usually called actuators and keep the laser beam on track and in focus by executing fine displacements. An additional servo motor is also used to perform large displacements of the laser spot along the disc radial direction. This electromechanical construction is usually designated as two-stage or sledge-actuator servo system.

The decoding electronics, also called channel decoder, processes the incoming high-frequency signal and regenerates the digital data stored on disc. It also performs error detection and correction, increasing therefore the reliability of data delivered further to the block decoder. Another function of the channel decoder is to regulate the speed of the turntable motor[6] according to a dedicated control algorithm.

The functionality of all electromechanical components is governed by a firmware program running in a microcontroller. However, in current CD-ROM drives, a common microcontroller is used to supervise the activities of both basic engine and data path.

---

[6]The term *spindle motor* is also commonly used. Notice that many drive architectures consider the turntable motor controller as part of the servo electronics and not of the channel decoder.

The role of the data path is to further process the digital signal output by the channel decoder and make it suitable for being used by the host system. The data path performs additional error detection and correction and, similarly to other disk-based storage devices, it also groups the read-out data into sectors. The communication between host system and drive is supervised from within the data path firmware.

Other electronic components, like semiconductor memories for storing firmware programs or power amplifiers for driving the motors are also needed. In addition to the above described building blocks, a CD-ROM drive may also contain a digital-to-analog (D/A) circuit and the related audio amplifiers to convert audio information (if audio discs are played back) into audible sound. These circuits are not shown in Fig.1.3. Finally, the disc eject mechanism (usually a tray) and its electromechanical control are not shown in this figure either, but they are equally important.

Chapter 2

The Optics

Since the disc itself is part of the optical subsystem of a CD-ROM drive, it will co-determine the way the recorded information is being processed. Fortunately, the disc has been standardized by the CD-DA Red Book [50,79] in 1981 and it has not been changed for more than 15 years.

2.1 The disc standard

A CD is a transparent plastic (polycarbonate) disc carrying a continuous spiral of impressed pits. The impressed surface is covered with a thin metallic layer, on top of which another plastic layer is being used for protection. The laser beam reads the profiled polycarbonate surface from below, perceiving therefore the impressed pits as bumps and being reflected back by the metallic coat. An intuitive representation of the disc read-out is depicted in Fig. 2.1-B.

The length of the pits varies, being dependent on the recorded data and on the modulation technique which will be further discussed in Section 4.6. The pits, having an approximate width of $0.6\,\mu$m, are arranged one after each other along a fictive continuous spiral called track. An enlarged view

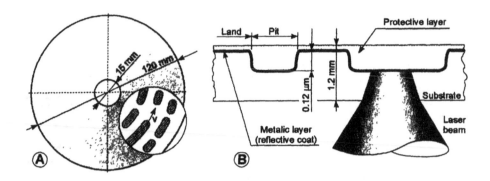

Figure 2.1   Disc impressed structure as seen from the substrate side (A) and in an enlarged cross-section along the disc spiral (B). The dimensions of both views are stylistically represented.

of the disc impressed surface is shown in Fig. 2.1-A. The spiral radius increases from the inner to the outer disc diameter while following the track in clockwise direction (viewed from the substrate side). The radial distance between tracks[1], called track pitch, remains constant all over the disc and may assume any value between 1.5 and 1.7 μm.

Another very important parameter of the pit is its geometrical depth. By choosing the right value for this parameter, the light reflected by a pit will approximately have an opposite phase[2] as compared to the incident beam. This gives rise to destructive interference, limiting therefore the amount of light coming back from the pits. On the other hand, any area surrounding a pit, usually called land, does not produce destructive interference and will consequently look brighter. By reading the intensity of the reflected light it will be therefore possible to detect the disc internal structure and to recover the recorded information. A detailed discussion of the disc read-out can be found in literature [11,12,13,46,115].

Apart from its sub-micron details, the external geometry of the disc has also been standardized. The same holds for the program area, where both

---

[1]Strictly speaking, the disc contains only one track along which the pits are impressed. However, the periodic structure arranged in the radial direction and determined by the disc spiral is usually indicated by the term *tracks*.

[2]For the purpose of optical recordings, the light can be described as an electromagnetic wave [1,9]. The wavefront points are characterized, among other parameters, by amplitude and phase.

the begin and the end diameters have specified dimensions, and for the spiral eccentricity[3].

Finally, there is one disc parameter having a significant importance for all high-speed CD-ROM drives: the linear density of recorded data along the disc spiral. According to the CD standard, the length of a pit/land may only assume nine discrete values (see further Section 4.6), while similar pits/lands have the same length at the inner as well as at the outer disc diameter. It follows that the relief structures are impressed with a constant linear velocity (CLV) along the disc spiral. This speed may only vary from disc to disc and the standards specify its numerical value between 1.2 m/s and 1.4 m/s. Although the read-out speed is not standardized, it has been generally accepted[4] to spin the disc at a CLV equal either to $v_a = 1.3 \pm 0.1$ m/s for audio playback or at $\mathcal{N}v_a$ for a data disc. The constant $\mathcal{N}$ is an integer number usually referred to as overspeed factor (or X-factor).

An overview of some standardized CD parameters is presented in Section 7.1. If necessary, extended information about the CD standards can further be found in [50,53,79,80].

2.2 The optical path

Although various optical arrangements are known for the read-out of compact discs, two of them are currently used on a large scale in CD-ROM drives. The optical paths associated with these two constructions are schematically depicted in Fig. 2.2. Basically, both constructions consist of a semiconductor laser, optical elements for guiding the laser beam, a photodetector to convert the incident light into electrical signals and, significantly to notice, the disc itself.

When compared with the construction **A**, the optical arrangement from Fig. 2.2-B uses a laser diode grating unit (LDGU) which incorporates the semiconductor laser, the photodetector (a set of photodiodes), and two diffraction gratings[5]. As an integrated optical component, the LDGU features functional stability, ease of assembly, and miniaturization [25,63].

---

[3]The eccentricity is defined as the radial offset (or run-out) between the center of the spiral and the center of the disc hole.

[4]High- and very high-speed CD-ROM drives do not follow this informal convention anymore. This subject will be discussed throughout Section 4.8.

[5]A diffraction grating is a periodic spatial structure whose physical dimensions are close to the wavelength of the incident light [9,59].

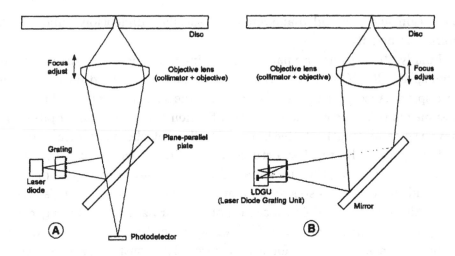

Figure 2.2   Two typical optical paths used in CD-ROM drives.

Moreover, the photodetector from within an LDGU does not need any separate optical adjustment during the assembly of the optical path itself.

Due to the purpose they serve, there are inherent similarities between the two optical constructions from Fig. 2.2. The semiconductor laser generates in both situations a light beam with a wavelength $\lambda = 780$ nm and an optical power of about 3 mW . This beam passes through a diffraction grating in order to split itself into a main beam, used for focus adjust and data read-out, and two secondary beams used for radial tracking.

One essential difference between the two presented optical paths is the use of a second grating element in the arrangement **B** to deflect the reflected beam towards the photodetector. The same role is played in the optical path from Fig. 2.2-A by the plane-parallel plate. The construction of the secondary grating will be further discussed in Section 2.5.

2.3 The laser spot

It is known from physical optics that a light beam passing through an aperture of dimensions which are small relative to the light wavelength will give rise to Fraunhofer or far-field diffraction [9]. This can be accomplished by arranging to have the source and observation points very far from the aperture. However, it can be shown mathematically [59] that an aperture

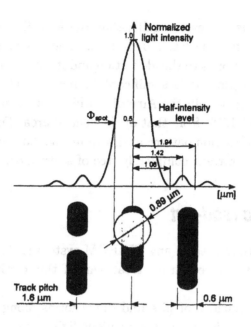

Figure 2.3   The laser spot and its light intensity.

situated just before a converging lens will also produce a similar Fraunhofer diffraction in the focal plane of the lens. The latter situation is also
characteristic for a CD-ROM optical path.

An important consequence of the far-field diffraction is the shaping of the
light intensity by maxima and minima which are given, in case of a circular
aperture, by a squared first-order Bessel function divided by a quarter of
its squared own argument [9,59]. Another consequence of the Fraunhofer
diffraction is the impossibility of having the laser spot focused to a perfect
point. The situation is graphically presented in Fig. 2.3. The central maximum of the light intensity corresponds to a bright circular region called the
Airy disk. This region is surrounded by alternate dark and bright rings as
given by the other minima and maxima. The radius of the Airy disk is

$$R_{Airy} = \frac{0.61\lambda}{NA} = 1.06 \ \mu m \tag{2.1}$$

where NA = 0.45 defines the numerical aperture of the objective lens from
Fig. 2.2 and λ = 780 nm. The numerical aperture as well as the laser
wavelength are specified in the CD standards [50,53,79,80].

The spot diameter can be defined either between the first two minima [1]
or full-width-at-half-maximum, usually denoted by FWHM [64,83]. The

effective spot size is, however, smaller than the Airy disk and for this reason we have also adopted the latter definition. The schematic representation from Fig. 2.3 corresponds to this definition and it can be shown that $\Phi_{spot} = 0.51\lambda/\mathrm{NA} = 0.89\ \mu\mathrm{m}$. It is significant to notice that 84 % of the total amount of light falls within the Airy disk while the intensity of the second maximum is about 1.7 % from that of the central area. Further, one should be aware that a diffraction pattern as drawn in Fig. 2.3 can only be obtained if the focused light wave is completely free of aberrations[6].

2.4 The disc read-out

The read-out method used in any CD-ROM system is that of the scanning microscope [12,13]. An extensive treatment of this method can be found in [46].

Basically, the sequence of pits and lands forms along the disc spiral a diffraction grating which splits the incident light into multiple diffraction orders [9,59], as depicted in Fig. 2.4. The disc read-out is accomplished by capturing the diffracted orders on a photodetector placed either behind or before the grating surface, situations which are referred to as transmissive and reflection read-out, respectively. The reflection read-out method is employed in all CD-based systems, including the CD-ROM drive.

After diffraction, the direction θ_m of a particular diffracted order m and the direction θ_0 of the incident light are related to each other through the grating equation

$$\sin\theta_m - \sin\theta_0 = \frac{m\lambda}{p} \tag{2.2}$$

with p denoting the grating period [9].

When carefully examining the disc impressed surface from Fig. 2.1-A, it appears that a two-dimensional grating is being present. The main one, which is responsible for the read-out method via the overlapping areas depicted in Fig. 2.4, is arranged along the disc spiral and consists of a long sequence of pits and lands of variable length. The smallest spatial

---

[6]On basis of geometrical optics, the aberrations can be described as failures of reflected or refracted light to give a point image of a point source. However, as the aberrations become small, they must be studied on the basis of diffraction theory [9]. In case of optical recordings, the aberrations (e.g. coma, astigmatism, distortion, etc.) are due to the geometry of spherical surfaces of lenses or mirrors.

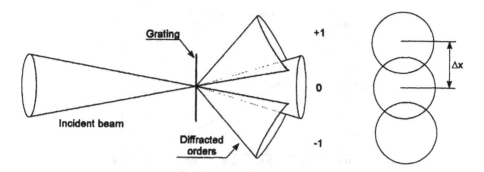

Figure 2.4 Zeroth and first diffraction orders generated in transmissive read-out by the pit/land grating structure, oriented along the disc spiral.

period of this grating is usually denoted by p and equals twice the shortest pit length. The other grating, which has a fixed period q equal to the track pitch, is arranged along the disc radial direction. It will generate another two overlapping areas which are perpendicular to the ones from Fig. 2.4 but are not represented anymore in this figure. This radial grating is responsible for a certain amount of cross-talk induced by the neighboring tracks into the read-out signal.

The read-out method employs, as mentioned above, only the grating aligned along the disc spiral. For the laser spot locally incident to the disc surface, the direction of this grating is perceived as being tangent to the spiral and moving with a given linear velocity $\mathcal{N} v_a$. The $\mathcal{N}$ and v_a parameters have already been defined in Section 2.1. The areas of the overlapping regions from Fig. 2.4 carry therefore time information related to the grating position. Also, because the pits (and lands) have variable length, the overlapping areas are affected by a spatial modulation.

When the whole zeroth order and both overlapping areas from Fig. 2.4 are captured by only one photodetector, the read-out method is called central aperture (CA) detection. This detection method is typically used in all CD-ROM systems. It can be shown [12] that the intensity of the light received by the CA photodetector is given by

$$I_{det}(t) \approx 2A_0^2 \left[1 + \left(\frac{A_1}{A_0} \right)^2 + \frac{2A_1}{A_0} \mathrm{MTF}(\Delta x) \cos \psi_{10} \cos \left(\frac{2\pi \mathcal{N} v_a}{p} t \right) \right] \quad (2.3)$$

where A_0 and A_1 are the amplitudes of the zeroth and respectively first diffracted order, ψ_{10} is the phase shift between these two light beams and

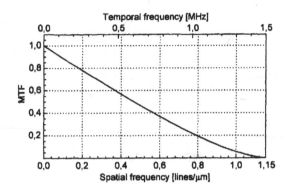

Figure 2.5   Modulation transfer function for central aperture detection.

$$\text{MTF}(\Delta x) = \frac{2}{\pi}\arccos\left(\frac{\Delta x}{2}\right) - \frac{\Delta x}{\pi}\sqrt{1 - \left(\frac{\Delta x}{2}\right)^2} \qquad (2.4)$$

is the modulation transfer function[7] (MTF). In case of CA detection, MTF is basically a measure of the cumulated overlapping areas depicted in Fig. 2.4 and depends on the spatial frequency $\Delta x = \lambda/(p \cdot \text{NA})$. MTF characterizes the low-pass filtering process undergone by the time-modulated third term from Equation (2.3). The modulation transfer function of a CD-ROM system with $\mathcal{N} = 1$ is plotted in Fig. 2.5. The corresponding cut-off spatial and temporal frequencies are equal to $2 \cdot \text{NA}/\lambda = 1.15 \ \mu\text{m}^{-1}$ and $2v_a \cdot \text{NA}/\lambda = 1.5$ MHz, respectively. The latter numerical value is obtained for a disc rotating at a linear velocity $v_a = 1.3$ m/s. Finally, it is important to mention that possible aberrations in the optical system will not affect the MTF but do introduce phase distortions in the read-out signal [40].

2.5   Optical error signals

It is of primary importance for any CD-ROM system to keep the laser spot in the middle of a particular track. Two error signals will therefore be needed for the servo electronics to automatically perform focus adjustment and radial tracking. Another remark concerns the shape of these error signals:

---

[7]The behavior of an optical system in the frequency domain [40] can be described by means of an optical transfer function (OTF). The modulus of OTF is known as modulation transfer function (MTF).

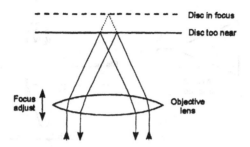

Figure 2.6   Beam focusing upon the disc impressed surface.

they have to monotonically increase (or decrease) when passing through zero, as a requirement for building a feedback control loop. It will be shown herein that both mentioned signals can optically be derived.

2.5.1 Focus error signal

When the scanning spot from Fig. 2.3 is perfectly in focus on the disc, the reflected laser beam is imaged back onto itself [12]. This optical phenomenon can better be understood from Fig. 2.6, where a disc out of focus is drawn, namely the objective lens is situated closer to the disc. Basically, it should always be possible to focus the light beam even if the disc does not contain any impressed structure at all.

A focus error signal can be obtained if some asymmetry is present in the optical path. Two methods are commonly used, which involve either the astigmatism deliberately introduced into the system or the separation of the light beam along its optical axis.

The principle schematically shown on Fig. 2.7 is known as the astigmatic method and is being based upon an optical aberration, called astigmatism. This distortion is introduced by cylindrical lenses or, in case of the optical path from Fig. 2.2-A, by the plane-parallel plate. An astigmatic image is also rotated with respect to its optical axis. As shown in Fig. 2.7, a focus error signal can be extracted if an arrangement of four detectors is being used. Details about the astigmatic method can be found in [12,83].

Other ways to generate the focus error are the single- and double-Foucault methods [12]. The former, also known from literature as the knife-edge method, is presented in Fig. 2.8. The asymmetrical beam is schematically drawn as being produced by a knife edge which obscures half of the beam. An out-of-focus situation will generate a mismatch between the light in-

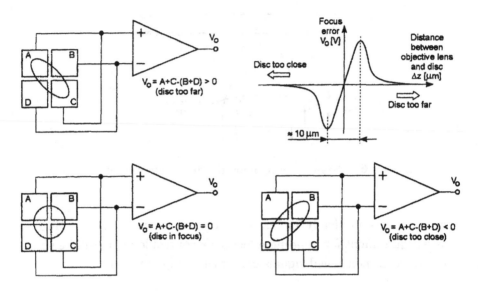

Figure 2.7   Typical arrangement and focus error signal for the astigmatic (4-detector) method.

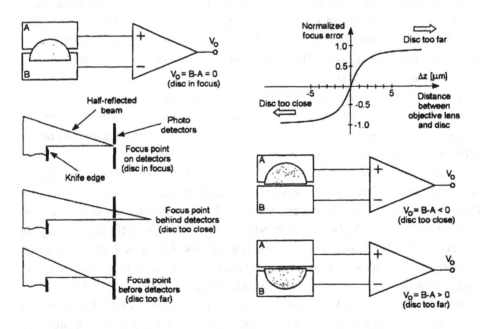

Figure 2.8   Generation of the focus error signal using the single-Foucault method.

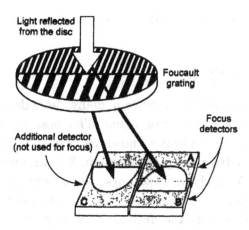

Figure 2.9   Foucault grating and its specific arrangement of detectors.

tensities perceived by the two detectors and consequently, a corresponding error signal can be derived.

The knife-edge method described above should be associated with the optical path from Fig. 2.2-B. However, instead of the knife edge, a holographic optical element (HOE) is used to introduce the necessary asymmetry along the optical axis [25,63]. Also called Foucault grating or simply hologram, the HOE has a special double structure as shown in Fig. 2.9 and divides the reflected laser beam in two halves which are separately directed towards the photodetector elements. One of these half-beams is focused onto a split detector and can therefore be used to generate a focus error signal. The other half is projected onto a third detector and, as it will be seen in Section 2.6, is necessary to generate the high-frequency signal whose information is carried by a complete bundle. The hologram, which is integrated within the LDGU, features high environmental stability and reduces the weight of the optical pickup unit (OPU), making it more suitable for fast access in high-speed CD-ROM systems.

Apart from the two methods described herein, the focus error can also be optically generated by means of a double-Foucault arrangement [12]. This method uses a prism instead of the knife edge to split the beam along its optical axis and two pairs of detectors to obtain the necessary error signal. Another known method is the spot-size detection. Also denoted in literature by the acronym SSD, the spot-size detection is similar to the double-Foucault method and employs a prism and two split detectors. While SSD can be found in some CD-ROM drives, the double-Foucault method has been replaced almost entirely by its simpler counterpart, the single-Foucault.

2.5.2 Radial error signal

Since the introduction of the CD system, many methods have been proposed for deriving an optical radial error signal. The most common strategies are described in [12] and are usually known as twin-spot radial detection (or 3-beam method), radial push-pull detection, 3-beam push-pull and respectively radial wobble method. Nowadays, the 3-beam method has become a standard in almost all CD-ROM drives. It uses two additional spots as shown in Fig. 2.10 to lock the desired track in between. This track will consequently be read out by the central spot which is equidistantly situated between the lateral ones.

The lateral spots, also called satellites, are derived from the initial laser beam by means of a grating structure (see also Fig. 2.2). The satellites are displaced aside from the central spot, both in the radial and tangential direction. The displacement in the radial direction equals a quarter of the track pitch and the grating should be properly adjusted by rotation in order to realize this distance. By using a separate detector for each satellite spot, a radial error signal can be created. The proper adjustment requires the detector signals being 180° out of phase with respect to each other.

As shown on Fig. 2.10, a sinusoidal signal will be generated while moving the spot along the disc diameter. The minima of this signal (i.e., when the light intensity drops to its lowest value) correspond to the positions of the considered spot just centered on a particular track. If one subtracts the two satellite detector signals, a DC-free radial error sinusoid

$$\mathrm{RE}(x) = \mathrm{SAT}_2(x, x_0) - \mathrm{SAT}_1(x, x_0)$$

$$= 2M \sin\left(\frac{2\pi x}{q}\right) \sin\left(\frac{2\pi x_0}{q}\right)\Bigg|_{x_0 = q/4} = 2M \sin\left(\frac{2\pi x}{q}\right) \quad (2.5)$$

can be obtained [12]. The requirement for a DC-free signal is to have both satellite spots of equal light intensities. The factor M defines the radial modulation and x represents the variable along the disc radius. The up-going slope of this error signal can consequently be used in a feedback control loop to keep the main spot centered above the right track.

We shall also mention at the end of this section that not all methods that can theoretically provide a radial error signal are equally suitable for

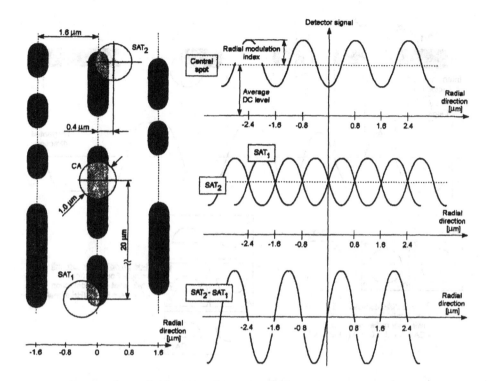

Figure 2.10 Generation of the radial error signal using the twin-spot method with a 3-spot optical pickup unit.

all types of CD systems. To give an example, a CD-Recordable cannot use the 3-beam method described above because the radial modulation index is almost zero if the disc is not recorded yet.

2.6 Generation of the HF signal

As the central spot from Fig. 2.10 remains locked to the desired track, the pit-land relief structure is passing by. The reflected main spot falls onto the central arrangement of detectors. As discussed in Section 2.5.1, a particular differential combination of these detectors is also used for the focus error. However, by gathering the information received by all central detectors, a high-frequency signal modulated by the disc relief structure can be derived. The modulation of this signal takes place in time as well as in amplitude, according to Equation (2.3).

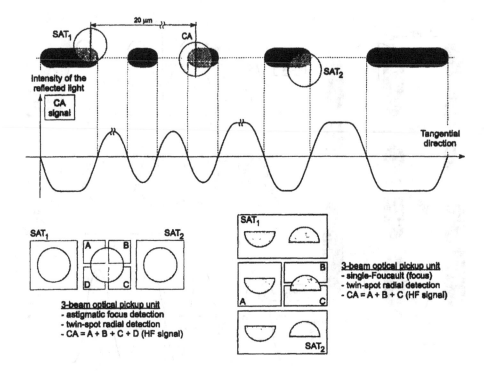

Figure 2.11   Generation of the high-frequency signal with a 3-beam optical pickup unit and the corresponding arrangement of detectors.

A schematic representation of the optically generated HF signal is depicted in Fig. 2.11 and the eye pattern[8] of this signal is shown in Fig. 2.12. The regions between the impressed pits (i.e., the lands) reflect the incident light without too much destructive interference and hence, the corresponding HF signal reaches maximum values. Conversely, a minimum level of the HF signal corresponds to a light beam which is strongly reduced by interference while reflected by a pit.

---

[8]The eye pattern can be obtained on an oscilloscope screen by superposing several time slices of the HF signal. The diamond-shaped space between the signal rising and falling edges is designated as the eye opening and provides various information about the signal channel (see also Section 4.4). A correct eye pattern can only be measured when the oscilloscope trigger level corresponds with a well-defined threshold, usually the DC component of the modulated measured signal. In telecommunications, this threshold or slicing level serves the regeneration of the transmitted clock.

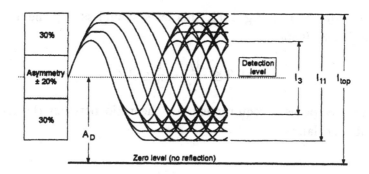

Figure 2.12   Ideal HF eye pattern.

The CA read-out signal as depicted in Fig. 2.11 is spatially modulated by the pit lengths, which are integer multiples of the channel bit[9]. The minimum and maximum fundamental frequencies of the modulation pattern (see further Section 4.6) are 196.4 and 720.3 kHz, respectively. As the length of a channel bit can be derived from the compact disc standards [50,53,79,80], all frequency components between 196.4 and 720.3 kHz may exhibit a slight variation of about ±8 %. The derivation of the modulation fundamental frequencies will be explained in Section 4.6.

The CD standards specify also two amplitude relations, namely

$$\frac{I_3}{I_{top}} = 0.3\ldots0.7 \tag{2.6}$$

$$\frac{I_{11}}{I_{top}} \geq 0.6 \tag{2.7}$$

which represent, practically, a provision for the noise possibly affecting the HF signal and for an optimal recovery of the recorded digital information. Both relations relate the amplitudes I_3 and I_{11} of the maximum and respectively minimum fundamental frequency components to the peak value I_{top} of the HF signal, before any high-pass filtering.

Further, it is known from telecommunication theory [21,26] that, in order for the system to recover the recorded information, a detection level A_D should be applied to the high-frequency signal. When related to the HF

---

[9]A channel bit carries the minimum amount of information which, for a digital signal, may only be either 0 or 1. When translated to the disc relief structure, the channel length represents a minimum physical length along the disc spiral. The subject will be discussed in Sections 4.6 and 7.3, and further details can be found in [12,46,64,70].

signal from Fig. 2.11 or Fig. 2.12, this level should satisfy the asymmetry condition [50,53,79,80]

$$-0.2 \leq \frac{A_D}{I_{11}} - \frac{1}{2} \leq 0.2 \tag{2.8}$$

which basically poses a requirement for the ratio between the lengths of similar pits and lands.

Chapter 3

The servo-mechanical subsystem

The CD-ROM servo electronics is basically associated with two control loops which keep the laser spot in focus on the impressed disc surface and, while playing back the recorded information, keep the laser spot following the disc spiral. In addition, the servo circuitry must also be able to find a particular location on the disc. This means crossing the tracks in the radial direction with high speed and without loosing focus, locating and locking the physical target track and resuming afterwards the playback state.

Strictly speaking, there are two more servo loops in a CD-ROM system, which regulate the speed of the turntable motor and load/unload the disc, respectively. The tray that carries the disc from the outside to the inside of the drive and vice versa is, in general, directly supervised by the basic engine microcontroller. A DC voltage with predefined time-dependent amplitude is applied to the tray motor until a certain condition is met (for instance, until a switch detects the end position of the tray). Other disc load/unload mechanisms[1] have similar operating principles. The turntable

---

[1] The so-called slot loader does only exhibit an opening in the front panel of the drive. When a disc is sufficiently inserted into the opening, it will be hooked by a soft-touch mechanism and pulled inside. The operation is reversed during unloading.

motor circuitry is usually built around the decoding electronics and will be separately described in Section 4.8.

We shall discuss within this chapter only the focus and radial servo loops, as being the two main feedback systems within a CD-ROM drive, apart from the one controlling the turntable motor.

3.1 Mechanical overview

The whole mechanical construction can be modeled from separate building blocks that interact with each other through elastic and damping elements. This is a common approach within the theory of automatic control [32,62].

Practically, almost all CD-ROM drives rely on mechanical concepts [88] similar to those depicted in Fig. 3.1. Two actuators, with their mobile parts totally suspended on elastic elements, are dedicated to keep the laser spot in focus and on track. The actuators can perform fine displacements along the normal and respectively radial direction relative to the disc while being positioned by a sledge at a raw radial location. The sledge forms a rigid body together with the turntable motor and the turntable itself, being further consolidated on what is called the subchassis.

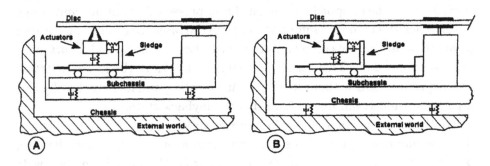

Figure 3.1   Schematic representation of the mechanical construction as seen from a servo point of view.

At this point, a difference can be distinguished between the constructions **A** and **B** from Fig. 3.1. The concept **A** has been successfully used for some years but, because of the relatively small mass elastically connected to the fixed world, it does not offer enough mechanical stability at high disc angular frequencies. The concept **B** combines the subchassis and the chassis itself in one body, providing therefore a heavy mass to passively damp the undesired vibrations which are generated by the rotating disc. For this reason, the latter construction is mostly used in very high-speed CD-ROM drives.

As a last remark, it is important to explain two commonly used abbreviations. OPU, which stands for optical pickup unit, defines the whole electromechanical ensemble (laser diode, lenses, photodetector and actuators) carried by the sledge. The compact disc mechanism (CDM) designates the subchassis itself together with the sledge, turntable motor, disc turntable, and OPU.

3.2 CD-ROM electromechanics

Any CD-ROM system is based upon five electromechanical actuators (motors) and their related control electronics, power drivers, gears, etc. The role played by each of these actuators is schematically depicted in Fig 3.2.

There are two rotary DC motors in the system, one of them spinning directly the turntable and the other one loading/unloading the disc. The latter motor may need additional mechanics to convert its rotational movement into linear (tray) displacement. Some portable CD-ROMs load the disc manually and do not have any dedicated load/unload mechanism.

A third rotary DC motor can be used to displace the sledge via a worm and gear combination. However, the very first CD-ROM drives did not use a

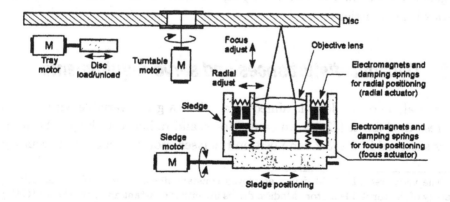

Figure 3.2 Schematic cross-section through the CD-ROM actuators and intuitive representation of all driving motors.

translation sledge but a swing-arm (similar to the one encountered in hard-disks) which involved pairs of magnets and electromagnets for positioning. Other constructions such as linear or step motors are also known but, for the time being, their use on a large scale is rather limited. Nevertheless, due to the importance of fast data access, the sledge electromechanics is getting increasing attention from the development engineers.

Finally, a CD-ROM drive needs two motors to correctly position the laser spot in focus on the disc impressed surface and radially along the desired track. These motors are usually designated as focus and radial actuator, respectively. They are able to carry out very fine displacements and rely on pairs of coils and permanent magnets which can move the objective lens in the vertical direction and radially across a number of tracks.

It will be also useful to indicate the magnitude of the movement that each electromechanical component is able to perform in current CD-ROM drives. The turntable motor rotates the disc with angular speeds between 5 and 120 Hz (i.e., between 300 and 7200 rpm), depending on the drive X-factor as well as on the spinning mode[2] of the disc. A tray or a slot-loading mechanism can load and unload the disc within 0.5 . . . 1.5 seconds and the sledge crosses the tracks in the radial direction with a maximum linear velocity of 200 . . . 350 mm/sec. The vertical displacements of the objective lens are typically of several hundreds of micrometers but, under certain circumstances[3] magnitudes of about 1 mm are not uncommon. The radial actuator can move the objective lens across a number of tracks but larger displacements, of 50 . . . 200 μm, are quite usual while playing back discs with eccentricity.

3.3   Dynamic disturbances and servo requirements

In general, a control loop is aimed to regulate a given variable within a set of prescribed limits [32,62]. In case of focus and radial servo loops, the actuators sketched in Fig 3.2 can position the laser spot with a certain accuracy

---

[2]The very first CD-ROM drives employed constant linear velocity (CLV) as driving strategy of the turntable motor. Modern drives use either constant angular velocity (CAV) or a predefined speed profile [103] between CLV and CAV. Concepts that combine two or even all of these three driving strategies are also known [105]. This turntable motor control will be discussed throughout Section 4.8.

[3]Discs with marginal specifications in the vertical direction (skew). Additional details are further given in Section 3.3.1.

along the normal and radial direction to the disc surface, respectively. The related control variable is, in both cases, the physical displacement.

In addition, large displacements are also performed by rapidly moving the sledge together with the carried actuators between two locations on the disc surface. It is quite common for the sledge servo to use the number of crossed tracks as controlled variable while the radial actuator is set to closely follow the fast sledge movement.

There are, however, error sources that disturb the positioning of the laser spot during the disc read-out. Similarly, the number of tracks crossed during a large sledge displacement can easily be miscounted due to various error factors.

The disturbances affecting the servo loops are either generated within the CD-ROM drive itself or are received from the external environment. Dynamically speaking, the control loop has to reduce any disturbance down to a value at which the disc read-out can still safely take place and, during a large sledge displacement, a target track can be fast and accurately located.

3.3.1 Disc vertical and track deviations

The standards [50,53,79,80] specify both the maximum deviations of the disc reflective surface along the focus direction and the radial deviations of the track. Disc irregularities up to these allowed maximum values should definitely not disturb the read-out process. The standardized deviations are summarized in Tables 3.1 and 3.2 and are defined as observed by the laser spot, while the disc is rotating at a scanning velocity $v_a = 1.2 \ldots 1.4$ m/s. For focus only, the tabulated values include the tolerances on substrate thickness, refractive index, and deflection. For radial deviations above 500 Hz, the standards specify a maximum tracking error which is due to noise within this frequency band.

3.3.2 Focus and radial servo requirements

It has been shown in literature [12,14,16] that a defocusing[4] of less than

$$\Delta z = \frac{\lambda}{2 \cdot \mathrm{NA}^2} \tag{3.1}$$

---

[4]Defocusing of the laser spot may have various causes. For instance, disc tilt (incorrect positioning of the disc on the turntable), disc skew and imperfections of the disc substrate lead to a periodical disturbance in the focus direction. The periodicity depends on the disc rotational frequency. Time-independent defocusing can also occur due to, for example, aberrations introduced by the objective lens.

Conditions	Parameter	Specification
Frequencies below 500 Hz	Maximum deviation from nominal position	± 0.5 mm
	Maximum RMS deviation	0.4 mm
	Maximum vertical acceleration	10 m/s^2
Frequencies above 500 Hz	Maximum deviation from nominal position	$\pm 1.0\,\mu$m

Note: The nominal position is defined with an ideal disc having a substrate thickness of 1.2 mm and a refractive index of the substrate equal to 1.55.

Table 3.1 Standardized vertical deviations of the information layer specified at the disc scanning velocity $v_a = 1.2\ldots1.4$ m/s.

Conditions	Parameter	Specification
Frequencies below 500 Hz	Maximum eccentricity of the track radius	$\pm 70\,\mu$m
	Maximum radial acceleration (eccentricity and unroundness)	0.4 m/s^2
Frequencies above 500 Hz	Maximum tracking error	$\pm 0.03\,\mu$m

Note: The eccentricity is specified relative to the inscribed inner circle of the center hole.

Table 3.2 Standardized radial deviations of the track specified at the disc scanning velocity $v_a = 1.2\ldots1.4$ m/s.

in each direction (towards and from the disc surface) is allowed while still preserving the reliability of the disc read-out. The above relation defines the total focal depth whose numerical value $\Delta z = 1.92$ μm is given by the CD-ROM parameters $\lambda = 780$ nm and NA $= 0.45$. When focusing the laser spot beyond Δz, the frequency response (MTF) given by Equation (2.4) becomes worse because of the non-uniform interference between zeroth and first diffracted orders [12]. The focus servo should therefore control the objective lens within $\pm\Delta z$.

Dynamically, a CD-ROM disc is specified while rotating at the constant linear velocity $v_a = 1.2\ldots1.4$ m/s, equal to the scanning speed of an audio disc[5]. The rotational frequencies corresponding to v_a lie between 3 and 9 Hz. It follows that a sensitivity[6]

$$\Delta S_{foc} = 20 \log \left(\frac{1.92 \ \mu\text{m}}{0.5 \ \text{mm}} \right) = -48 \ \text{dB} \tag{3.2}$$

is needed in the focus loop at these low frequencies, where the maximum vertical deviation from Table 3.1 has been taken into account. It should be noticed that Equation (3.2) gives only the sensitivity necessary for the rejection of those disturbances introduced by the disc itself. However, other dynamic disturbances that contribute to defocusing are usually much smaller and can be neglected under normal circumstances. The only exception occurs while playing back a heavily unbalanced disc and this situation will be discussed later, in Section 3.3.3.

In case of radial tracking, a similar approach can be considered. The scanning spot, which is supposed to closely follow the read-out track, might slightly deviate from its ideal position while still remaining within a given tracking error. This error is mostly determined by the total amount of cross-talk introduced into the HF read-out signal by the neighboring tracks.

It can be shown that, due to cross-talk considerations, radial deviations of $\pm 0.2 \ \mu\text{m}$ can easily be tolerated in a CD system [12]. The necessary sensitivity of the radial control loop becomes

$$\Delta S_{rad} = 20 \log \left(\frac{0.2 \ \mu\text{m}}{70 \ \mu\text{m}} \right) = -51 \ \text{dB} \tag{3.3}$$

when calculated at the audio scanning velocity v_a. Again, only the disc itself as disturbance source has been considered in Equation (3.3).

An intuitive representation [70] of the disc specifications in both focus and radial directions is depicted in Fig. 3.3. The plots are given for a disc rotating at the constant linear velocity v_a. The corner frequencies can easily be derived using the equation $a_{max} = 4\pi^2 A f^2$ with a_{max} and A representing the maximum acceleration and deviation, respectively[7]. When the disc rotates at higher speeds, say $\mathcal{N} v_a$, the corner frequencies of both plotted specifications must linearly be shifted by the overspeed factor $\mathcal{N}$. The focus and radial accelerations, on the other hand, will increase by the factor $\mathcal{N}^2$

---

[5]See Sections 2.1 and 4.8.
[6]The sensitivity of a closed control loop will be further discussed in Section 3.4.
[7]For example, the lowest corner frequency for the focus specification is equal to $\sqrt{10/(0.5 \cdot 10^{-3})}/(2\pi) = 22.5$ Hz.

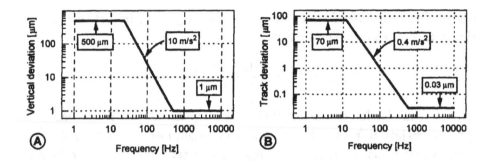

Figure 3.3   Graphical representation of the disc specifications in focus
(A) and radial direction (B), for a disc scanning velocity
$v_a = 1.2 \dots 1.4$ m/s.

when measured along the ordinate of these plots. The whole disturbance
spectrum in focus or radial direction should lie below the respective plotted
specification for any given frequency f.

As a final remark, we notice that the disc deviations given in Tables 3.1
and 3.2 exceed the practical values which characterize the current CD-ROM
discs. This affirmation is especially valid for the focus and radial maximum
accelerations and leads to more relaxed bandwidth specifications of the respective control loops.

3.3.3 Additional internal disturbances

Apart from the deviations introduced along the focus and radial directions
by the disc itself, the read-out process is also jeopardized by other error
sources from inside the CD-ROM drive.

It is significant to mention that not only the vertical and radial directions
are important for the spot positioning, but the tangential direction should
also be taken into account. Errors along this direction can be due to some
displacements of the light source, vibrations of tilted mirrors but also play
and misalignment in the sledge mechanics. Fortunately, these types of errors
do not affect the read-out performance of a CD-ROM disc to such an extent
that a servo correction becomes necessary [12]. The only optical systems
which do care for tangential errors are those needing an accurate time base,
such as the Laser Vision (CD Video).

Looking again at the radial direction, a very important disturbance is
the eccentricity introduced by the shaft of the turntable motor, turntable

itself, and the positioning of the disc on the turntable. They add up to the maximum run-out of the track radius from Table 3.2 and lead to a maximum eccentricity of about 125 μm. The radial servo loop should therefore be calculated for a total reduction

$$\Delta S_{rad} = 20 \log \left(\frac{0.2 \ \mu\text{m}}{125 \ \mu\text{m}} \right) = -56 \text{ dB} \qquad (3.4)$$

at the scanning velocity $v_a = 1.2 \ldots 1.4$ m/s.

Another very important disturbance source in the system is the disc itself as a rigid mass in rotation. The disc is not a perfect geometrical body and consequently, its rotation will generate a centrifugal force which is proportional to the distance ΔR_{gr} between the center of gravity and the rotation axis. The amplitude of the unbalance force is given by

$$F_{unbal} = m(2\pi f_{rot})^2 \Delta R_{gr} \qquad (3.5)$$

where m is the disc mass and f_{rot} is the rotational frequency of the disc. Practical values for this force lie between 4 and 40 mN when measured at $f_{rot} = 10$ Hz.

The effect of the unbalance force is perceived inside the drive as vibrations of the whole subchassis. It is the disc center of gravity rotating at a rate given by f_{rot} that determines the frequency of the subchassis oscillations. The actuators, which are elastically coupled to their carriage (see Fig. 3.1), will tend to follow the subchassis oscillations and hence give rise to focus and radial errors. These errors can be very large, reaching under extreme conditions values of tens of micrometers. It is possible to partially attenuate these errors by properly designing the focus and radial servo loops. However, the nonlinear components of these errors can only be dealt with by using, for example, very specific control techniques. An example of a nonlinear effect introduced by F_{unbal} is the amplification of the play between sledge and its guiding mechanics. This play remains practically unobservable if a well-balanced disc is read out but otherwise, it can hardly be compensated for. In addition, if a track-counting mechanism is used during an access procedure, it can also be disturbed by F_{unbal}.

In order for the heavily unbalanced discs to be played back, some mechanical measures should be taken. Adding more mass to the subchassis as in Fig. 3.1-B when compared to Fig. 3.1-A can improve the situation. Also, the mechanical tolerances can be tightened or even self-balancing turntables can be used. Another solution employs a dedicated algorithm to detect the unbalance of the played-back disc. Depending on the degree of unbalance, the disc rotational speed might be lowered to a particular value at which the CD-ROM functions are not disturbed anymore.

A secondary effect induced by unbalance forces is the propagation of the internally-generated vibrations to the outside world. Usually, a CD-ROM drive is mounted on the mechanical frame of its host system (PC) which means that a rigid link is established between the two mechanical ensembles. The PC industry and especially the hard-disk manufacturers can only tolerate a well-defined disturbance environment for their relatively quiet devices. As a consequence, the CD-ROM drive manufacturers are the ones called to adopt specific measures to reduce the spectrum and the amount of mechanical vibrations.

3.3.4 External disturbances

Other sources of disturbances which can affect the disc read-out process are situated outside the player cabinet. They are generated by the external physical environment and do introduce errors within the focus and radial servo loops. Some examples include the car navigation systems using a CD data base and the computer systems with optical storage devices working in hostile environments (e.g. those devised for the military equipment).

However, as the rotational speed of the disc has been continuously increased over the years, the external disturbances became of less importance when compared with those generated by the disc unbalance. More attention has therefore been paid to suppress the latter disturbances, unless the drive must be specifically designed to accommodate extreme environmental conditions.

3.4 Servo loops for focus and radial adjustment

It has already been mentioned in Section 3.3 that both focus and radial servo loops use physical displacement as control variable. Each of these loops is therefore regulating the linear position of the objective lens and employs an optoelectronic detector to generate the error signal and several electromechanical components. The block diagram of a position servo loop is illustrated in Fig. 3.4.

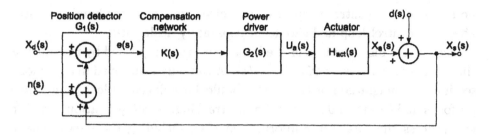

Figure 3.4   Block diagram of a position-control servo loop.

3.4.1 Position control loop

The spot position, denoted in Laplace[8] domain by $X_s(s)$, is determined by the displacement of the objective lens and should coincide at any moment in time with a predefined reference $X_d(s)$ situated on the disc itself. This reference is given by the disc reflective layer in case of the focus control loop and by the center of the read-out track in case of the radial loop. Clearly, the laser spot must follow the disc deviations discussed in Section 3.3.1 while disturbances $d(s)$ originating   inside[9] and outside the CD-ROM drive will influence its controlled position.

The error signal $e(s)$ is optically generated as already discussed in Section 2.5 and converted into an electrical signal by a photodetector. The whole arrangement used for error generation, including some electronics needed to read the photocurrents,[10] is depicted as a single building block and designated as position detector. In general, a closed control loop is also affected by the measurement noise $n(s)$ which is introduced while reading the feedback signal [32]. For our particular case of focus and radial control, $n(s)$ has both a DC component or offset and a component which is randomly distributed in the frequency domain (white noise). The measurement noise originates mostly in the photodetector itself and the related electronics. Fortunately, as it will be further discussed in Section 5.2, the random component of $n(s)$ can be sufficiently minimized such that it can

---

[8]The complex Laplace variable is designated by $s = j\omega$ with $j = \sqrt{-1}$ and ω being the angular frequency of interest.

[9]These disturbances include not only those internally generated as discussed in Section 3.3.3, but also the sledge contributions in a two-stage (sledge-actuator) radial loop. The block diagram of a sledge-actuator radial loop is depicted in Fig. 3.9.

[10]See Sections 5.2 and 5.3 for further details.

be neglected for control purposes. The offset, on the other hand, will still affect the control loop and should be separately dealt with.

The correct displacement of the laser spot is accomplished by minimizing the tracking error $e_t(s) = X_d(s) - X_s(s)$. As already mentioned in a previous section, the magnitude of this error should be reduced to less than 2 μm for focus and less than 0.2 μm for radial tracking. A compensation network will process the total error information $e(s)$, including the measurement noise, and will actuate the objective lens towards the desired position. The transfer functions of the position detector, compensation network, actuator, and power driver of the actuator have been denoted by $G_1(s)$, $K(s)$, $H_{act}(s)$, and $G_2(s)$, respectively.

Depending on the signal of interest, the feedback loop from Fig. 3.4 can be described by any of the equations

$$X_s(s) = \frac{H(s)}{1 + H(s)}\left[X_d(s) - n(s)\right] + \frac{1}{1 + H(s)}d(s) \qquad (3.6)$$

$$e_t(s) = \frac{1}{1 + H(s)}\left[X_d(s) - d(s)\right] + \frac{H(s)}{1 + H(s)}n(s) \qquad (3.7)$$

where

$$H(s) = G_1(s)K(s)G_2(s)H_{act}(s) \qquad (3.8)$$

designates the open-loop transfer function. As indicated by (3.6), the sensitivity $S(s) = 1/[1 + H(s)]$ needs being small for disturbance rejection. The same requirement is also needed in (3.7) to minimize the tracking error $e_t(s)$. The magnitude of $S(s)$ at low-frequencies should in particular cover the numerical values from Equations (3.2) for focus and (3.4) for the radial loop. On the other hand, a good rejection of the measurement noise takes place when the magnitude of the complementary sensitivity $1 - S(s) = H(s)/[1 + H(s)]$ approaches zero. The compromise between disturbance rejection, noise rejection, and small tracking error is accomplished by the compensation network which makes $|S(s)|$ and $|1 - S(s)|$ small at low and high frequencies, respectively.

Detailed information related to feedback control can be found, for example, in [32,61,62]. We shall approach here the actuator and the compensation network, while the position detector will be separately discussed in Section 5.2. As a remark, notice that $G_1(s)$ and $G_2(s)$ may be approximated in most cases by real gains within the operating frequency range (up to several kHz) of the focus/radial loop.

3.4.2 Focus and radial actuators

The two tiny motors used to vertically and radially position the laser spot
are usually called actuators. The displacement directions of the objective
lens which are associated with the focus and radial actuators have already
been indicated in Fig 3.2.

Each actuator uses a separate pair of coils and permanent magnets to de-
velop an electrical force along the desired direction. The permanent magnets
are mechanically fixed while the coils are attached to the moving objective
lens. This construction is known in literature as a linear voice-coil motor
and is widely used in other disk storage devices [66]. The voice-coil motor
benefits from a relatively small moving mass which is specifically aimed to
improve the step response of the position loop as well as to reduce the power
dissipation.

The moving part of both actuators, including the objective lens, is suspended on elastic elements that are designed to separate as much as possible
the two directions of movement. However, a certain amount of cross-talk
is present between focus and radial loops which are, otherwise, completely
independent from a control point of view. The cross-talk is due to mechanical and electrical coupling through the objective lens and electrical fields
of the coils, respectively. At this stage, a careful electromechanical design
should be considered to reduce the mutual focus-radial cross-talk to values
acceptable for both loops.

A simplified linear Laplace model of the focus/radial actuator is presented
in Fig. 3.5. Two equations govern the dynamics of the actuator, namely

$$u_a(t) = Ri(t) + L\frac{di(t)}{dt} + u_b(t) \tag{3.9}$$

describing its electrical behavior and

$$F(t) = M_{act}\frac{d^2x(t)}{dt^2} + D\frac{dx(t)}{dt} + Kx(t) \tag{3.10}$$

characterizing its mechanics. The notations used here are $u_a(t)$, $u_b(t)$ and
$i(t)$ for the applied voltage, back emf and respectively current through the
actuator coil, and R, L for the coil resistance and inductance. Further, M_{act}
is the actuator moving mass, D is the damping constant (viscous friction
coefficient) of the surrounding medium, and K is the constant characterizing
the elastic elements. The time-dependent linear displacement $x(t)$ should
be regarded as relative to a reference position of the objective lens from
Fig 3.2, which generally corresponds to $i(t) = 0$.

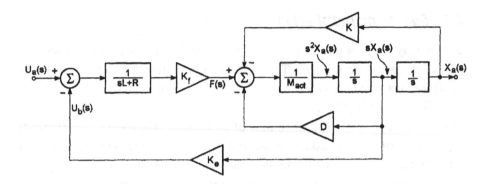

Figure 3.5   Block diagram of an ideal CD-ROM focus/radial actuator.

The two above equations are linked through the electrical force $F(t) = K_f i(t)$ where K_f denotes the force constant. When K_f and the back emf constant K_e from Fig. 3.5 are expressed in SI units, they are defined by the same numerical value.

The actuator transfer function can easily be derived from Equations (3.9) and (3.10). Together with the relation $F(t) = K_f i(t)$ they lead to

$$H_{act}(s) = \frac{K_f}{sL + R} \cdot \frac{1}{M_{act}s^2 + \left(\dfrac{K_f K_e}{sL + R} + D \right)s + K} \tag{3.11}$$

where an electrical time constant $\tau_{el} = L/R$ can be immediately identified. Typical values for a CD-ROM actuator are $R = 10 \ldots 20\ \Omega$, $L \approx 0.1$ mH, $K_f = 0.2 \ldots 0.6$ N/A, $M_{act} = 0.4 \ldots 0.8$ g, $D \approx 0.01$ Ns/m, and $K = 20 \ldots 100$ N/m. As mentioned before, the back-emf constant expressed in V/m has the same numerical value as K_f in N/A.

Further, it can also be seen from Equation (3.11) that a resonance frequency characterizes the actuator transfer characteristic. At low frequencies, when only the mechanical part of this equation is taken into account (i.e., $K_e = 0$), the natural undamped frequency f_n and the damping ratio ξ become

$$f_n = \frac{1}{2\pi}\sqrt{\frac{K}{M_{act}}} \tag{3.12}$$

$$\xi = \frac{D}{2\sqrt{M_{act}K}} \tag{3.13}$$

The resonance peak of a second-order system with respect to the DC gain is

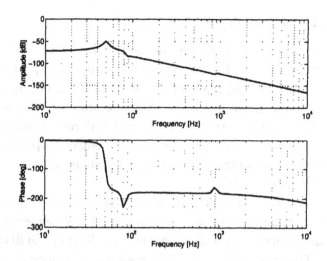

Figure 3.6   Bode plots of an actuator with three resonance modes.

given by the quality factor $Q = 0.5/\xi$. Additional details on these subjects of control theory can be found in [32,62] and a general approach of the CD actuator is presented in [12].

Finally, some remarks must be made with respect to the *parasitic* resonances of the actuator. Due to its mechanical construction, there are always some vibration modes which disturb the transfer characteristic (3.11) and add undesired resonance peaks. These modes can be modeled as second-order signal paths in parallel with the mechanical part of the ideal actuator from Fig. 3.5. The Bode plots[11] of a real actuator with $f_n = 50$ Hz and two more parasitic modes at about 85 and 900 Hz, respectively, are shown in Fig. 3.6. It is not always possible to eliminate these resonances and neither to shift them at higher frequencies where they could possible be reduced by the loop attenuation. Moreover, these parasitic modes are usually temperature-dependent and exhibit variations from one actuator to another. In some high-speed CD-ROM systems, the disc rotational frequencies are lowered when servo difficulties arise at higher ambient temperatures and especially in combination with heavily unbalanced discs.

---

[11]The Bode plots represent a graphical tool for studying the frequency characteristics of a given linear system. If the complex transfer function of such a system is denoted, for example, by $H(s)$ with $s = j\omega$, the Bode plots depict the amplitude $|H(s)|$ and phase $\arg\{H(s)\}$ as functions of the operating angular frequency ω or frequency $f = \omega/(2\pi)$ of the system.

3.4.3 The compensation network

The compensation network should be designed to meet several goals or design criteria within the frequency range of interest: (i) minimization of the tracking error $e_t(s)$, (ii) rejection of the disturbances $d(s)$, (iii) rejection of the measurement noise $n(s)$, (iv) dynamic stability, (v) prevention of signal saturation at the actuator input, and (vi) robustness with respect to system tolerances. These goals regard a control loop working within its linear region. The control process itself is usually denoted by the term *tracking* (of the input signal).

The first and second design criteria mentioned above require the open-loop transfer function $H(s)$ to have a large gain at low frequencies where the first harmonic of any periodic disturbance as well as of the disc deviations is situated. This particular operating region of the focus/radial loop is determined by the disc angular frequency $\omega_d = \mathcal{N}v_a/(2\pi r)$, with r being the radius of the read-out position.[12]

For noise rejection, on the other hand, the gain of $H(s)$ should be kept low at high frequencies. Finally, the dynamic stability can be achieved by satisfying the gain and phase margin requirements[13] in the middle of the operating frequency region. Detailed information related to these classical subjects on automatic control can further be found in [32,61,62].

One of the most known techniques used to compensate for a second-order plant with own natural frequency is the addition of a proportional, integral and derivative (PID) controller, eventually followed by a low-pass filter (LPF). This compensation network is described by the equation

$$K(s) = \left(K_p + \frac{K_i}{s} + sK_d \right) \frac{1}{\tau s + 1} \tag{3.14}$$

where K_p, K_i, and K_d are the gains associated with each of the three PID actions and τ is the time constant of the LPF. A Bode plot of such a compensation network is given in Fig. 3.7. At low frequencies, the integral part of the controller increases the loop gain to achieve the necessary disturbance rejection. The proportional part brings also additional gain into

---

[12]For $v_a = 1.2\ldots1.4$ m/s, the low-frequency region extends between 3.3 and 9 Hz and scales linearly with $\mathcal{N}$ at higher scanning velocities.

[13]A closed-loop system remains stable if the argument of $H(s)$ does not decrease below $-180°$ at the point $s = j\omega$ where $|H(s)| = 1$. The gain margin gives the factor by which the loop gain may be increased before the system becomes unstable, i.e., where $\arg\{H(s)\} = -180°$. The phase margin is defined by the relation $\varphi_m = 180° + \arg\{H(j2\pi f_c)\}$ where f_c designates the cross-over frequency at which $|H(j2\pi f_c)| = 1$.

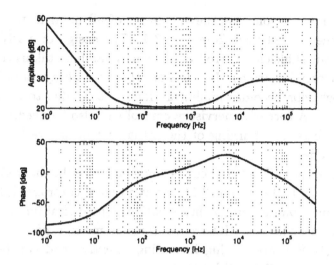

Figure 3.7 Bode plots of a typical PID controller, followed by a low-pass filter.

the system, shifting upwards the whole amplitude plot from Fig. 3.6 and increasing hence the loop bandwidth.[14] The differential part of the compensator plays its role around the system cross-over frequency, where the actuator mass can hardly follow the relatively fast excitations. At these frequencies, the denominator from Equation (3.11) is mostly determined by the second-order term $M_{act}s^2$ because $\tau_{el} = L/R$ is usually situated far outside the bandwidth. The differential action sK_d will therefore reduce the order of the open-loop transfer function $H(s)$ to one (in the neighborhood of the bandwidth), which is a necessary condition for system stability. Finally, the low-pass filter takes care of the rejection of the measurement noise at high frequencies.

Some other effects introduced by a PID controller are also present but they are less obvious, although not less important, and follow only from typical calculations [61,62]. For instance, due to the proportional-integral part, the rise and settling times are penalized. The proportional-derivative part, on the other hand, will add some damping to the system whereas the steady-state response is not affected.

---

[14] The bandwidth of a control system [32] is defined as the frequency at which the signal power drops to half of its DC value. When related to the closed-loop transfer function, this definition implies a magnitude drop equal to 3 dB with respect to $H(j0)/[1+H(j0)]$.

Another remark concerns the tolerances of all physical parameters involved in the control loop from Fig. 3.4. The design of a PID regulator should provide sufficient control robustness by carefully taking into account these parameter variations. In addition, there are always situations when the loop behavior becomes nonlinear (for instance, due to a heavy occasional shock). A sort of supervisory control will also be needed to prevent undesired effects, like burning of the actuator coil, and restore the linear operation mode.

Practically, the PID network can be implemented either in analog domain, as already given by Equation (3.14), or as a digital controller [61] based on a discrete-time counterpart of this equation. The digital implementations, in the form of dedicated PID architectures or multipurpose digital signal processors (DSPs), are getting increasing attention and can be found in most of the current CD-ROM drives.

It is significant to mention at the end of this section some other techniques which have been proposed to control the focus and/or radial position loop. An approach which is equivalent to a PID compensator is that of a state estimator in combination with linear state feedback [66]. QFT controllers which are based on quantitative feedback theory [23], the H_∞ control [32,107,108] or learning schemes which help suppressing the repetitive disturbances [69] have also been studied. All these control techniques have the same global objectives already mentioned at the beginning of this section but they do employ specific methods for improving the loop performance. However, because the PID controller remains very efficient when compared to its implementation costs, the market of servo ICs has hardly been penetrated by any other type of regulator.

3.5 Focus control

The laser beam is kept in focus by a circuit which incorporates a position loop and a nonlinear control unit (NCU). A block diagram of this circuit is presented in Fig. 3.8.

The PID regulator, which is assumed herein to regulate the focus position, performs only within a linear operating region as already discussed in the previous section. However, controlling the focus loop does not represent always a linear process. The system has to be initialized and started up, the focus error should be calibrated and eventual offsets need being elimi-

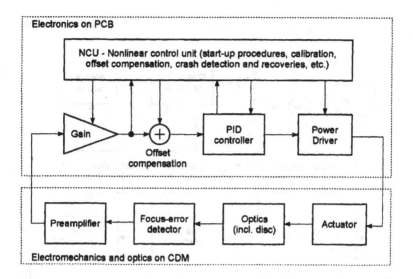

Figure 3.8   Block diagram of the focus control.

nated. Moreover, the functionality of the position loop should be continuously monitored to detect possible crashes and initiate the proper correction algorithms to recover the lost focus. Such crashes can occur due to, for instance, discs with thickness closed to specifications, in combination with heavy unbalance and surface scratches. The task of the NCU is to assist the PID compensator during various nonlinear operations and, if needed, take over the focus control.

In general, the nonlinear control unit is a complex circuit based on hardware and software. The decisions taken by the focus NCU during operation are also dependent on information received from both radial loop and turntable motor control. It becomes therefore quite difficult to separate the focus NCU functions from those of other loops. One example is the detection of a very damaged (too many scratches) or very unbalanced disc, for example, which cannot be played back at nominal speed. As a consequence, the drive specifications can be relaxed by reducing the motor rotational frequency.

3.6 Radial control

Due to the large disc radial dimensions relative to the track pitch, the spot position is controlled by a two-stage electromechanical ensemble, as already

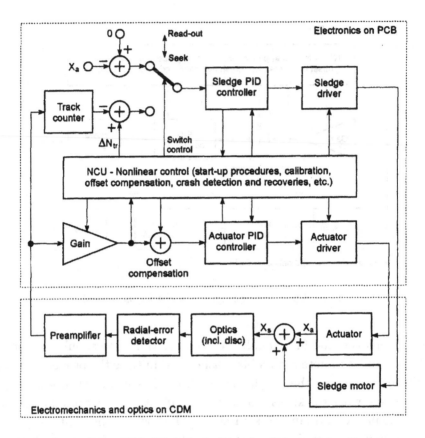

Figure 3.9   Simplified block diagram of the radial control.

discussed in Section 3.2. While the very first CD-ROM drives featured a swing-arm to access any desired track on the disc, the current drives use only carriages with linear displacement for the same purpose.

A general block diagram of the radial control is depicted in Fig. 3.9. The actuator branch, which is similar to the focus loop described throughout the previous section, is controlled by a PID network during its linear operation. Typical Bode plots for a PID-based radial open-loop are given in Fig. 3.10. In addition to the actuator displacement, the carriage (sledge) movement brings its own contribution to the positioning of the laser spot along the disc radius. At this point, a distinction should be made between on-track positioning, also called track following, and the radial seek displacement.

The nonlinear control unit (NCU) is also present in the radial servo loop. As with focus control, this unit takes care of all initialization and start-up

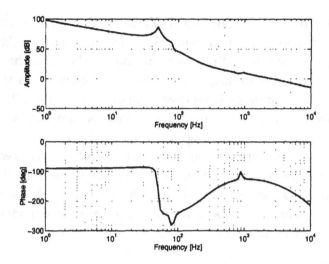

Figure 3.10   Typical open-loop Bode plots for the PID-regulated
actuator section of the radial control.

procedures, calibrations, crash detections and recoveries. The crash situations, however, are of a more complex nature than those encountered in a focus loop. Not only the on-track radial control can fail but crashes can also occur during a seek action, while the sledge is moving with high velocities across the disc. Practically, most of the problems affecting the on-track and seek control are due to discs with heavy defects, strongly unbalanced, or are caused by external shocks.

3.6.1 On-track radial control

While the CD-ROM system delivers data through its host interface, the laser spot must obviously follow a given track. At a higher system level, this operation mode is usually called playback or data read-out. At a lower level, this function is designated by the term *track following* and represents a particular state of the radial servo circuitry.

Basically, only the actuator loop from Fig. 3.9 is responsible for the accurate spot positioning, while the sledge is set to slowly follow the actuator movements. This configuration requires a closed sledge loop of relatively low bandwidth[15] and controlled by its own compensation network (a PID

---

[15]The bandwidth of the sledge control loop during radial tracking amounts for several hundreds Hz, depending also on the scanning velocity of the disc.

regulator, for instance). The signal X_a to be tracked is selected with a seek/read-out switch and represents the actuator position relative to the sledge, as given reference. While continuously at playback, the spot moves towards the outer disc radius, leaving the sledge behind. As the displacement range of the actuator is relatively small (see Section 3.2), the sledge will also have to advance but at a slower pace, without following the fast tracking movements of the laser spot.

A measure of the actuator position with respect to the sledge can be obtained, for example, by reading the integrator value from the actuator PID controller. Other methods employ the reading of the actuator back emf by means of an additional coil or using a state space model to predict the actuator movements [66].

3.6.2 Radial seek control

In general, for a disk-based storage system, the servo is said to perform a seek when the radial control is directed to place the read head on a track different from the present one [66]. At a higher system level, like the host interface, the term of *data access* is also commonly used and points indirectly to a seek procedure.

A universal control algorithm able to perform the desired seek action can easily be explained with the help of Fig. 3.9. Before issuing a seek command, the system microcontroller has to calculate the number of tracks ΔN_{tr} to be radially crossed. The formula used for this calculation is

$$\Delta N_{tr} = \frac{1}{q} \left(\sqrt{\frac{D_i^2}{4} + \frac{v_a q S_{final}}{\pi}} - \sqrt{\frac{D_i^2}{4} + \frac{v_a q S_{init}}{\pi}} \right) \qquad (3.15)$$

where q is the track pitch, D_i is the inner diameter of the program area, v_a is the linear velocity of the recorded data and S_{init}, S_{final} are the initial and final subcode[16] timing before and after seek, respectively. The seek procedure is also referred to as a jump action. Depending on the sign of ΔN_{tr}, the seek will be outward- or inward-oriented.

The displacement of the laser spot during a seek action should be fast and reliable. In general, a counting mechanism which increments for each

---

[16]The subcode timing defines the position of a data cluster along the disc spiral. Because the linear velocity v_a of the recorded data remains constant all over the disc, this position can be indicated in time units, starting with 0 seconds at the beginning of the spiral (see further Section 6.1). The subcode timing increases strict-monotonically in steps of 1/75 seconds from the inside to the outside of the spiral.

track crossed in the radial direction is necessary. The radial error signal from Fig. 2.10 can easily be used by this mechanism to accumulate track-crossing information.

The sledge loop, which played a secondary role during track following, takes now control of the seek action. Although not explicitly shown in Fig. 3.9, the actuator is set to follow the fast sledge displacement[17] while the track-counting circuit provides the necessary feedback information. The sledge controller acts to increase the number of crossed tracks towards the value given by Equation (3.15). However, just before the end of the sledge displacement, an algorithm is necessary to switch the radial control from sledge back to the actuator loop and smoothly provide the track acquisition. This algorithm is always needed, irrespective of the type of control (not only PID) being used.

3.7   Focus and radial control in high-speed drives

In high- and very high-speed CD-ROM drives, the frequency spectrum of all disturbances and disc deviations scales linearly with the overspeed $\mathcal{N}$ (recall that $\mathcal{N} v_a = \omega_d r$, with r being the disc radius at the current read-out point).

It can be shown [101] that, when a PID controller is used to regulate the position of the laser spot, the bandwidth of both focus and radial servo loops must be increased by the factor $\mathcal{N}$ to achieve the same accurate positioning as at 1X. The very first 1X CD-ROM drives used a radial/focus bandwidth of about 500 Hz and it follows that 16 kHz would be needed at 32X. However, it turns out in practice that smaller bandwidths are still effective because the accelerations that must be reduced are situated below those derived from disc specifications (see Section 3.3.1). This favorable situation is primarily due to the improvements that took place in the disc manufacturing process since the compact disc has been standardized.

On the other hand, the high unbalance forces (see Section 3.3.3) would also require larger bandwidths of both focus and radial loops. As these forces generate vibrations perceived by the outside world, it is first desired to reduce them to acceptable values. In this respect, the CDM mass is

---

[17]One third of the distance between the inner and outer disc radii is typically used to measure the seek performance and is called third-stroke (see also Section 7.4.2). Current CD-ROM drives are able to execute a third-stroke sledge displacement (i.e., 11 mm) within 50...100 ms.

increased as shown in Fig. 3.1-B or the disc speed is set to a lower value. In both cases, however, the bandwidth servo requirements can be relaxed to several kHz.

Chapter 4

The decoding circuitry

Along with the optics and servomechanics, the decoding circuitry plays one of the main roles within a CD-ROM system. Also designated as channel decoder, the decoding electronics processes the read-out HF signal, regenerating the digital data embedded into the disc relief structure.

4.1 The CD system as a communication channel

The high-frequency signal depicted in Fig. 2.11 forms the starting point towards the understanding of the decoder functionality. This subject is, however, not strictly related to CD-ROM data processing but it has its roots deep into the theory of digital communications [21,26].

The compact disc can be considered as a transmission medium for the carried signal. A digital convention will be agreed upon such that, for instance, the light reflected by lands will be denoted by a 1 (one). Conversely, the light reflected by pits (negative peaks of the HF signal) will bear the digital 0 (zero) notation. The disc becomes therefore the carrier of a digital signal. Usually, a data stream will have to be encoded first and modulated before being sent along a communication channel. The CD-ROM system

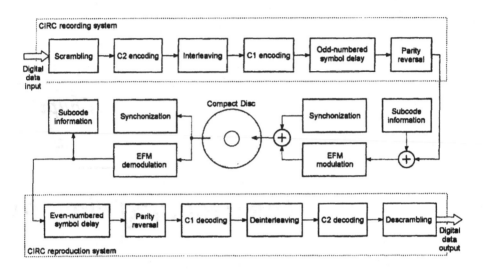

Figure 4.1   Schematic representation of the CD data channel.

does also obey this general rule and the associated communication channel [70] is schematically drawn in Fig. 4.1.

The function of each block represented in Fig. 4.1 will be explained throughout the next sections. There are two global algorithms having an influence upon the performance of the CD channel: the modulation scheme, called Eight-to-Fourteen Modulation (EFM) and the error detection and correction scheme, called Cross-Interleaved Reed-Solomon Code (CIRC).

Apart from the EFM and CIRC algorithms employed in all types of CD-based drives, the CD-ROM data undergoes additional scrambling and error-correction coding which have not been shown in Fig. 4.1. The corresponding descrambling and error detection and correction operations take place at the drive data path level and they will be discussed throughout Chapter 6. The CD-ROM basic engine is, however, only responsible for the extraction of the digital data stream from the transmission medium (disc). The two necessary operations performed by the channel decoder are only those related to EFM demodulation and CIRC error detection and correction, respectively.

4.2 Data-recovery general architecture

When analyzed one level deeper than a transmission channel, the decoding circuitry reconstructs the data stream based on several general concepts.

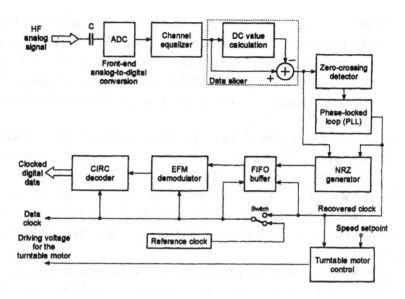

Figure 4.2   Basic data-recovery circuits arranged as for a fully-digital CD-ROM channel decoder.

They are schematically shown in Fig. 4.2 and are, again, common to the world of digital communications [26].

First of all, somewhere along the decoder path, the analog signal should be converted into digital domain. Although some of the decoding operations (channel equalization, for instance) can be performed in a straightforward manner on an analog signal, the current trend is to work as much as possible with digital signals. When such an approach is used, an analog-to-digital converter (ADC) precedes the whole channel decoder as shown in Fig. 4.2.

A second general concept common for these types of circuits is the channel equalizer. In case of the CD-ROM system, the equalizer compensates for the low-pass character of the optics, namely for the modulation transfer function (MTF) plotted in Fig. 2.5.

Further, the separation between the positive and negative peaks of the HF signal can be achieved by determining the points where this signal intersects a certain threshold (slicing) level. In the CD-ROM system, the modulation of the HF signal is DC free and hence, the intersection points are also called zero-crossings. The slicing level can be obtained by dynamically calculating the DC value of the incoming signal. The circuit used to determine this threshold level is usually referred to as data slicer [26] or automatic threshold control [54].

The fourth dedicated circuit encountered in all communication channels, either analog or digital, is the phase-locked loop (PLL). The task of the PLL is to regenerate the clock from the transmitted signal. This clock can subsequently be used to reconstruct the transmitted stream of digital symbols. Also, the regenerated clock serves as time base for further signal processing.

Finally, the last two general decoding concepts regard the demodulation of the EFM signal and the CIRC detection and correction of the possible errors. A first-in-first-out (FIFO) buffer may also be used to regulate the data flow in some CD-ROM drives[1] or while playing back audio discs. However, it is possible to bypass the FIFO buffer by setting the same clock for its input and output data streams. This feature is used in most high-speed CD-ROM drives.

Apart from typical decoding functions, a disk-based transmission channel needs a control circuit to spin the turntable motor (also designated as spindle motor). The driving voltage for this motor is generated by comparing a given speed reference with the real disc velocity, the latter being proportional with the recovered clock.

The data-recovery circuits depicted in Fig. 4.2 will be separately approached throughout the next sections. However, the analog-to-digital converter [82] which has become a standard function in almost any electronic equipment will not be dealt with.

4.3 The channel equalizer

As any transmission medium has its own transfer characteristics, the carried signal suffers amplitude and phase distortion. In case of the CD-ROM channel, the optics (including the transmission medium, i.e., the compact disc) has already been optimized [12,70] but its intrinsic low-pass behavior requires a separate compensation. This behavior is given by the modulation transfer function (MTF) already shown in Fig. 2.5.

From the viewpoint of physical optics, the low-pass characteristic of the MTF is due to the limited resolving power of the lenses. Some optical compensation methods have been proposed in literature [51] but the electronic compensation is much cheaper and effective. It only employs a high-pass fil-

---

[1]A FIFO buffer is only needed in CD-ROM drives based on constant linear velocity (CLV) motor control. The subject will be further discussed throughout Section 4.8.

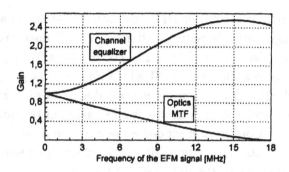

Figure 4.3   Typical amplitude response of a CD-ROM channel equalizer and the MTF of the transmission channel, for a 12X drive.

ter that boosts the high-frequency components of the read-out signal while exhibiting a constant delay in the frequency band of interest. A typical amplitude plot of such an equalizer is given in Fig. 4.3 along with the modulation transfer function of central aperture (CA) detection.

A number of algorithms are known for designing the channel equalizer and, among them, the Nyquist methods [26] or the raised-cosine roll-off filtering [26,51] are widely discussed in literature. For transmission signals used in optical recordings, and in particular for the EFM-based CD-ROM signal, the specific modulation techniques which are employed allow the designer of the equalizer to deviate from the standard Nyquist methods [6].

Most channel equalizers are aimed to reduce the intersymbol interference[2] (ISI). However, they can also be optimized for improving the signal-to-noise ratio (SNR) of the received signal [6]. While the very first equalizers used in CD-ROM systems were designed for one or two fixed data bit rates, the situation has been totally changed since the dramatic increase of the X-factor. The current CD-ROM drives are delivering data within a large range of bit rates which would basically require an equalization function adaptively depending on the data throughput.

---

[2]The intersymbol interference (ISI) is due to the finite bandwidth of any communication system. A signal consisting of abrupt pulses will be filtered improperly as it passes through the communication channel and the pulses, corresponding to transmitted symbols, may be smeared into adjacent time slots. The resulted waveform does not display a clear separation between neighboring symbols, impeding therefore the subsequent data detection.

Finally, we notice a counter effect introduced by the channel equalizer: despite its design towards reducing the ISI, it will also steepen the edges of the HF signal depicted in Fig. 2.11. For a *theoretically-perfect* read-out signal, this leads to an increased ISI which implies a larger jitter of the re-generated clock [12]. On the contrary, optical aberrations or the defocusing of the laser spot, which contribute to the distortion of the analog HF signal, are alleviated by a properly designed equalizer. As a perfect and continuous read-out spot can hardly be realized, the channel equalizer does always bring an improvement to the data recovery process.

4.4   *Signal slicing and zero-crossing detection*

The continuous sequence of pits and lands carried by the compact disc, as a transmission medium, represents a serial-bit line code which is generally designated in literature as nonreturn-to-zero (NRZ) signaling [21,26,83].

Following the disc read-out, the recorded pit/land NRZ sequence becomes, in the analog domain, the HF signal (see also Fig. 2.11). As a convention, the binary value 1 is associated with the positive peaks resulted from land reflections while the binary 0 is assigned to the negative peaks (pit reflections).

In order to recover the recorded digital sequence, a decision level should be found as shown in Fig. 4.4. This threshold is usually referred to as slicing level and the associated electronics is called slicer. The term of regenerative repeater is also commonly used in digital communications to denote the slicing circuitry.

The analog HF signal undergoes, as already discussed in Sections 4.3, a low-pass distortion due to the MTF but also channel equalization. Basically, the quality of the incoming HF signal can be appreciated by looking at its eye pattern. For example, the ideal (distortionless and without ISI) eye pattern from Fig. 2.12 can be compared with a distorted one plotted in Fig. 4.5. The eye opening provides very valuable information [26] such as: (i) timing error allowed for clock recovery, given by the width inside the eye; (ii) sensitivity to the slicing level, given by the slope of the open eye; and (iii) noise margin, given by the height of the eye.

In order for the digital NRZ sequence to be recovered without errors, the slicing circuitry should detect the horizontal signal level passing through the middle of the eye opening. The detection of this threshold level in a CD-

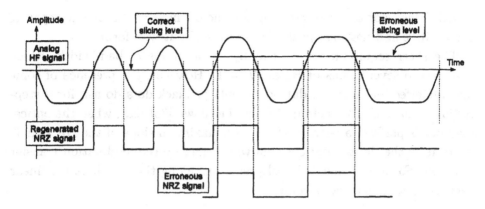

Figure 4.4   The slicing level and its influence upon the regeneration of the NRZ signal.

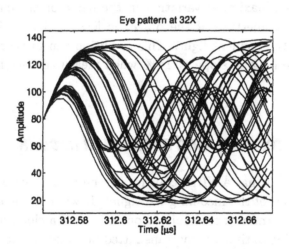

Figure 4.5   Distorted eye pattern of the HF signal (simulation results).

ROM system is facilitated by the modulation (EFM) of the recorded digital data which produces a DC-free HF signal[3] The modulation technique will be discussed in Section 4.6. A relatively simple slicer as previously drawn in Fig. 4.2 can therefore be employed. It will only have to dynamically calculate the DC level of the signal and subtract it from the equalized HF

---

[3]It is assumed that all DC offsets introduced by the preprocessing electronics are removed by the coupling capacitor depicted in Fig. 4.2.

signal. This leads to a zero-crossing threshold which can be used to separate the positive and negative peaks from the read-out waveform.

The slicer requirements are usually set in terms of bandwidth, which amounts to several tens of Hz at $\mathcal{N} = 1$. However, the presence of some disc imperfections such as data drop-outs or black dots do ask for a step-response analysis of this circuit. It can be shown [65] that, when the optical read-out is perfect (aberration-free), the maximum slope of the HF signal[4] is given by the limited optical bandwidth, i.e., by the modulation transfer function. For a disc that is locally read out, along the spiral, at the linear velocity $\mathcal{N}v_a$, this slope becomes

$$\left. \frac{di(t)}{dt} \right|_{max} = 1.6\, I_{pp} \cdot \frac{\text{NA}}{\lambda} \cdot \mathcal{N}v_a \tag{4.1}$$

where I_{pp} is the maximum variation in the detector current[5], NA $= 0.45$ is the numerical aperture of the objective lens, and $\lambda = 780$ nm is the wavelength of the incident light. This condition can further be translated into a maximum signal slope of 14 % when measured over half of the eye width.

4.5 Clock recovery and associated circuits

It has been shown that a NRZ sequence can be reconstructed by detecting the zero-crossings of the incoming HF signal. It will only be possible to un-ambiguously determine a transmitted one or zero if a clock signal indicates the exact position in time of any transmitted bit. Usually, no separate clock channel is used in digital communications but the time base is embedded into the carried signal and needs being regenerated at the receiving side.

The regenerated clock should have a very precise frequency and phase relationship with respect to the incoming HF pulses. The process of ex-tracting the clock from the transmitted signal is called bit synchronization. A clarification is however necessary, regarding two dedicated terms: bit and frame synchronization, respectively. While the latter is used to mark and

---

[4]This maximum slope is obtained at a zero-crossing, between two adjacent longest land and pit (or vice versa).

[5]The maximum variation in the detector current corresponds to the difference between 100 % and 0 % reflection levels.

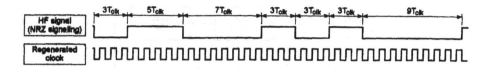

Figure 4.6 Transmitted clock regenerated from the HF signal after slicing and zero-crossing detection.

distinguish groups[6] of data, bit synchronization serves separating the bit intervals from each other.

The complexity of the bit synchronizer depends on the sync properties of the digital signaling [26]. The ideal synchronous detection takes always place, however, when the clock pulses are generated at the time moments corresponding to the maximum vertical eye opening. In CD-ROM systems, a phase-locked loop (PLL) circuit has become a standard solution for clock recovery. A schematic representation of the NRZ digital signal and the corresponding regenerated clock is presented in Fig. 4.6.

A basic rule which should be obeyed in most communications systems, and in particular within a CD-ROM system, concerns the duration of the transmitted pulses. Neither a one nor a zero signaling is allowed to exceed a given maximum time length. The reason for this condition is related to the PLL capability to keep generating the clock by looking only at the transitions between two NRZ levels. If the distance in time between these transitions becomes too large, there will be no timing information available for the PLL. In case of the CD-ROM system (and disk-based storage devices, in general), this constraint is resolved by modulating the digital data stream before being recorded on disc (see further Section 4.6).

As already mentioned in Section 4.2, the clock regeneration can also take place in the analog domain. The current trend is, however, to use all-digital PLLs [6,7,90] which can be completely adjusted and reprogrammed while in normal operation. Independent of its implementation, a phase-locked loop relies on three main building blocks as depicted in Fig. 4.7. A linear-gain phase detector determines the phase difference between the incoming NRZ signal and the clock generated by the PLL itself. The loop filter will provide the necessary dynamic stability and improve the tracking error. A

---

[6]In general, one should talk about block synchronization. This operation, which separates the words, frames or data packets from each other, can be accomplished by embedding predefined patterns into the data stream [6].

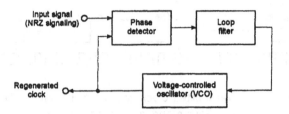

Figure 4.7   Block diagram of a phase-locked loop.

proportional-integral (PI) filter is commonly used to reduce the tracking error when frequency steps are applied at the PLL input. The regenerated clock is produced by a voltage-controlled oscillator (VCO) driven by the filtered phase difference. The counterpart of the VCO in all-digital PLLs is the discrete-time oscillator[7] (DTO) whose periodic output waveform is usually a step-wise sawtooth.

In addition to the phase loop depicted in Fig. 4.7, the clock regeneration may also employ a frequency loop to help the pull-in process of the PLL. When the frequency difference between the clock embedded within the incoming NRZ signal and the regenerated clock is significant, the phase acquisition cannot take place. The frequency loop regulates the VCO output towards a frequency value situated within the pull-in range of the PLL, improving the behavior of the phase capture. Such situations occur, for example, while starting up the CD-ROM drive or while repositioning the laser spot to another track (i.e., during a seek procedure). Valuable information related to all-digital clock recovery can be found in [6].

Three issues are of paramount importance for the PLL behavior. First, the capture and lock ranges, as key parameters, determine the pull-in and the tracking performance, respectively. The capture performance can also be improved by using additional (non-)linear circuits, the so-called aided acquisition [68]. The second remark concerns the stability of the PLL as a control system. In this respect, a proper design of the loop filter should be taken into account. Finally, the jitter[8] reduction [6,99] and the behavior

---

[7]The terms digitally- or numerically-controlled oscillator are also used.

[8]The jitter cumulates statistically all small time variations of a repetitive signal, within a given frequency range, reflecting therefore the quality of the whole transmission channel. For example, the cumulated small variations exhibited by the period of a transmitted digital signal is usually designated as data-to-data jitter. Similarly, the jitter can be measured between the edges of digital data and recovered clock, respectively (clock-to-data jitter).

of the PLL in the presence of noise [6,36,58] are very important for correct bit detection. The PLL bandwidth which is necessary for clock recovery at $\mathcal{N} = 1$ is about 3 kHz.

4.6 EFM technique, signal demodulation and frame construction

It has already been mentioned in Section 2.1 that the length of a pit (or land) does not change from the inside to the outside of the disc, when measured along the disc spiral. These fixed spatial lengths correspond to time intervals measured between the edges of the NRZ data stream and lead to a constant data throughput when the scanning velocity along the spiral is kept unchanged.

The discrete length values which a pit/land can assume are determined by the modulation technique being used. Basically, the goal of any modulation scheme is to convert the existent information in such a way that it falls within a frequency band convenient for the transmission channel and can be recovered at the receiver in a stable manner [26]. The CD-ROM system employs the Eight-to-Fourteen Modulation (EFM), which belongs to the class of run-length limited (RLL) codes [93].

An RLL code describes a sequence of binary data where upper and lower limits are imposed for the number of consecutive zeros and ones[9]. These limits are given by two parameters, designated as d and k, and the RLL sequence is said to be a (d, k)-constrained code. The d-constraint forbids the occurrence of less than d consecutive zeros between two logical ones. It follows that d controls the highest transition frequency, reducing therefore the intersymbol interference (ISI) when the binary signal is transmitted over a bandwidth-limited channel. The maximum length of an all-zero sequence is limited by the k-constraint, which controls thus the low-frequency content of the data stream. This parameter provides an adequate rate of transitions between the two binary levels 0 and 1, which consequently allows the PLL to regenerate the embedded bit clock. The parameters d and k are chosen depending on various factors such as channel response, desired data rate (and information density, in case of disk-based systems), jitter, and noise characteristics [93,54]. The RLL data streams are characterized by bit sequences of minimum and maximum lengths equal to $d + 1$ and $k + 1$, respectively.

---

[9]It is assumed that 0 and 1 are the logical levels on which the binary sequence relies.

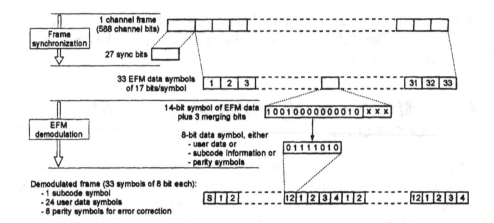

Figure 4.8   EFM demodulation of a channel data frame.

EFM is derived from a $(2, 10)$-constrained code and was designed to fit the frequency response of the optical system [70,77,96]. EFM is an efficient and highly structured RLL code which, from a performance point of view, is very tolerant to disc imperfections, increases the data density on the disc by 25 %, and enables a stable clock recovery [44,83,93]. According to the $(2, 10)$-constraints, the minimum and maximum RLL sequences (i.e., 100 and 10000000000) correspond to 3 and 11 channel bits, respectively. When converted into NRZ signaling, any 1 from the RLL sequence marks a transition between the two NRZ logical levels (see also Fig. 4.4). In terms of physical lengths on the disc, the 3- and 11-bit patterns correspond to the shortest and longest pit/land, respectively.

By using a fixed mapping (look-up table), EFM translates any 8-bit symbol from the original data stream into 14 channel bits. There are 267 words of 14 bits which obey the $(2, 10)$-constraints but only 256 words out of 267 are used for the translation table. However, not any two 14-bit words can be transmitted one after the other because of the minimum/maximum runlength violations. For this reason, 3 merging bits are added after each 14-bit sequence. Although 2 merging bits would be enough to satisfy the RLL requirements, a third bit helps reducing the low-frequency content of the digital stream. When this strategy is applied, the noise in the servo-band frequencies (< 20 kHz) is suppressed by about 10 dB [93].

A schematic representation of the demodulation process and the various constituents of a data frame are given in Fig. 4.8. Apart from the parity symbols which will be discussed in Section 4.7, the frame also contains a subcode symbol and a 27-bit synchronization pattern[10] The subcode symbols from 98 consecutive[11] demodulated frames are cumulated to provide various user information (for instance, the subcode timing along the disc spiral, which can be used to access the desired data). Details on the frame construction can be found in the CD standards [50,53,79,80] or in literature [44,70].

It is also useful to present within this section an overview of the bit rates in a CD-ROM system. The starting point for these calculations is the rate of 7350 frames/second[12] at 1X linear speed, which originates in the CD-DA standard. The results are shown in Table 4.1.

Bit stream	Number of bits per frame	Bit rate [Mb/s]
User data	$(12 + 12) \times 8 = 192$	1.4112
Raw data	$(1 + 2 \times 12 + 2 \times 4) \times 8 = 264$	1.9404
Channel data	$(1 + 2 \times 12 + 2 \times 4) \times 17 + 27 = 588$	4.3218

Table 4.1   Various bit rates within a CD-ROM system.

Another interesting overview is related to the frequency components generated by the read-out of the discrete pit/land lengths. In this case, the starting point is the channel rate of 4.3218 Mb/s. The highest frequency of the RLL transitions which corresponds to the minimum pit/land length equals 720.3 kHz. For any EFM pattern kT with $k = 3, \ldots, 11$ the associated frequency is given by

$$f_{kT} = \frac{4.3218 \cdot 10^6}{2k} \cdot \mathcal{N} \tag{4.2}$$

---

[10]The sync pattern consists of two consecutive maximum-length sequences (10000000000) followed by 10xxx, where 'xxx' denotes three merging bits.

[11]See further Section 6.1.

[12]For an audio disc, the demodulated frame from Fig. 4.8 contains two streams of 12 symbols, for the left and right channel, respectively. Each audio sample contains 16 bits and hence, one frame carries $12 \times 8/16 = 6$ samples per audio channel. As the sampling rate of the audio signal equals 44.1 kHz (see Section 5.4), there are $44.1 \cdot 10^3/6 = 7350$ demodulated frames needed to generate one second of music. In case of the CD-ROM system, the 2×12 symbols simply contain computer data instead of sampled audio, preserving therefore the number of 7350 frames/second.

with $\mathcal{N}$ being the overspeed factor of the read-out data. The physical length of the corresponding pit/land profile is

$$L_{kT} = k \cdot \frac{v_a}{4.3218 \cdot 10^6} \qquad (4.3)$$

where $v_a = 1.2 \ldots 1.4$ m/s is the linear velocity of the recorded information. The length of the shortest pit/land is $L_{3T} = 902.4 \pm 69.4$ nm and, accordingly, the frequencies given by Equation (4.2) may also exhibit variations.

4.7 CIRC error detection and correction

The demodulated frame, as depicted in Fig. 4.8, incorporates eight parity symbols which are arranged in two equally-sized groups. It should be recalled that any demodulated symbol contains 8 bits. The parity symbols carry special information needed for the detection and correction of erroneous bytes that might be present in the read-out frame.

Due to either defects in the manufacturing process or damage of the disc surface during use (scratches, fingerprints, dust), the read-out digital signal may not match the original source signal used to create the disc [20,48]. As an advantage of digital communications, the reliability and performance of the transmission (or storage) system can be increased by detecting and correcting a *certain amount* of received errors.

Two types of errors can be identified within the read-out signal: those which are randomly distributed among the individual bits and the burst errors [21]. The latter can be spread over hundreds and even thousands of data bits. In general, the integrity of data [83] can be quantified with three parameters: bit error rate (BER), block error rate (BLER) and burst error length (BERL). BER, which specifies only the number of errors but not their distribution, is defined at the receiver side as the ratio between the number of erroneous bits and the total number of bits received. BLER measures the number of blocks (frames) of data per second that have at least one erroneous bit, and BERL counts the numbers of consecutive data blocks (frames) in error.

The CD-ROM system makes use of the Cross-Interleaved Reed-Solomon Code (CIRC) which has been developed by Sony and Philips for CD-DA. Valuable information related to CIRC can be found in [12,45,70]. The Reed-Solomon (RS) codes employed in CIRC belong to the class of linear block codes[13] and are very efficient in correcting burst errors [117]. If redundant information is provided in the form of parity symbols, a rather simple algorithm can be used to detect and correct a certain number of erroneous or even missing symbols within the corresponding received code word. The calculation of the parity symbols relies on specific rules from linear algebra[14] and resumes, finally, to a linear combination of all data symbols. A good practical approach of the error correction techniques can be found in [77,83] and an extended mathematical treatment of the coding theory is given, for example, in [24,78].

Before proceeding further, two important results from the theory of RS codes should be mentioned. First, it will be always possible to correct t symbols if $2t$ parity symbols are added to the original data. Second, if erasures[15] are already known (i.e., detected by a decoder-independent device), it will be possible to rebuild at most $d_m - 1$ erroneous symbols, where d_m is the minimum distance[16] associated with a given code.

The two Reed-Solomon codes employed in CIRC are called C_1 and C_2 and are both based on 8-bit symbols (bytes). A codeword of C_1 contains 28 data symbols and uses 4 additional symbols for parity calculations. Accordingly, C_1 is designated as a (32,28) RS code. Using a similar notation, C_2 is a (28,24) RS code, i.e., with codewords of 24 data and 4 parity symbols, respectively. It follows that both C_1 and C_2 have a minimum distance $d_m = 5$ and can therefore correct directly at most two symbols (within one codeword) or at most four erasures. However, it is also possible to simultaneously perform t straightforward corrections and e erasure corrections provided that $2t + e \leq d_m$. The parity symbols of C_1 and C_2 are usually denoted by P_i and Q_i, respectively, with $i = 1 \ldots 4$.

---

[13]A block code uses a sequence of n symbols (called codeword), each symbol comprising m bits. Only k symbols out of n carry user data, while the other $n - k$ symbols contain parity information. A linear code has the property that any two codewords of n symbols can be added and their (modulo-2) sum represents a valid codeword too.

[14]Details on this subject are beyond the scope of this book. However, it is important to mention the notation GF(q) which is encountered in all CD standards and designates a Galois field of q elements [78]. The Galois fields are needed for the description of Reed-Solomon codes and their associated encoding/decoding techniques.

[15]When a received symbol is unreliable (but not necessarily in error), its known position within the codeword is called erasure.

[16]The distance d between any two codewords is given by the number of places in which these codewords differ from each other. The minimum distance d_m of a code is the smallest value of the distance among all possible combinations of two codewords. For an (n, k) RS code, $d_m = n - k + 1$ and t errors can be corrected if $2t + 1 \leq d_m$. However, erasure correction is more efficient, since $d_m - 1$ erasure corrections can be performed.

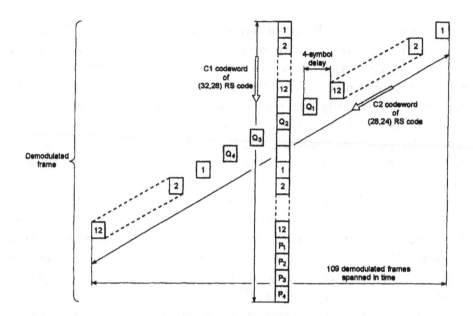

Figure 4.9   Construction of the C_1 and C_2 codewords based on symbols from demodulated frames.

An intuitive construction of the C_1 and C_2 codewords based on symbols from demodulated frames is presented in Fig. 4.9, and the block diagram of a CIRC decoder is depicted in Fig. 4.10. The subcode symbols are not represented in these figures because they undergo a separate detection and correction process which will be discussed in Section 6.1.

The received demodulated frame enters the CIRC decoder with all symbols at the same time, in parallel. The data decoding begins with a 1-symbol delay performed upon all even-numbered symbols. Two frames are therefore needed at this stage to rebuild an original undelayed 28-symbol frame. These delays, introduced during the encoding process, improve the correction of small burst errors by spreading two adjacent corrupt symbols over two user frames.

The next operation reverses the polarity of the parity symbols in a bit-wise manner. The polarity reversal took previously place in the encoder in order to prevent producing all-zero codewords at the receiving side[17] It

---

[17] All-zero codewords can be obtained, for instance, if the HF signal vanishes due to an interruption along its path. An all-zero codeword will be validated by the decoder as being without errors.

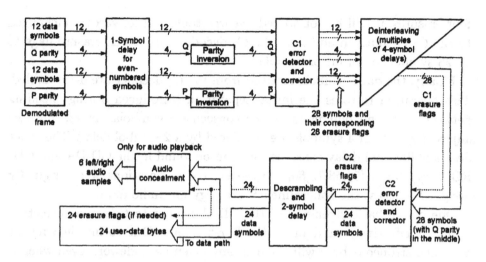

Figure 4.10 Block diagram of the CIRC error detection and correction.

becomes therefore possible to detect bit insertions and deletions caused by clocking difficulties [117].

The very first correction operation is performed by the C_1 decoder. It uses the P-parity symbols to correct at most two data symbols out of 32. Different detection/correction strategies can be used [77] provided that equation $2t + e \leq 4$ remains valid. However, the strategy adopted in current CD-ROM drives is to mark all symbols from the C_1 codeword as erasures and perform no correction if there are more than two erroneous symbols. The erasure flags are accompanying their symbols towards the C_2 decoder.

Following the C_1 correction, the deinterleaving is taking place. Originally, 27 symbols out of 28 were delayed during encoding, in order, 4,8,12 and up to 108 frames, respectively. The deinterleaving process, which restores the original symbol positions, needs 109 demodulated frames to generate a C_2 codeword (see Fig. 4.9). The possible burst errors are hence spread over many C_2 codewords which facilitates the error detection and correction at C_2 level. Significantly to mention that, along with any symbol, its associated erasure flag attached by C_1 decoder is also deinterleaved.

The next operation is carried out by the C_2 decoder which uses the Q-parity symbols as well as the erasure flags provided by C_1 to accomplish its task. Although the inequality $2t + e \leq 4$ allows for several possible decoding strategies [2,3,77], most current CD-ROM drives perform four erasure corrections at the C_2 level. If the incoming codeword contains more than four erasures, it will be considered uncorrectable and forwarded unchanged to the C_2 output. In addition, the C_2 decoder will attach erasure flags to all 28 symbols of an uncorrectable codeword.

Finally, the data symbols received from the C_2 decoder are descrambled. This operation restores the initial separation introduced during encoding between even and odd pairs of two consecutive symbols. Next, any two adjacent groups of 4 symbols are separated by a 2-symbol delay. The exact descrambling and delay constructions can be found in the CD-DA and CD-ROM standards [50,53,79,80]. However, both operations were originally designed to assist the concealment[18] strategy for audio data.

One of the basic strengths of the CIRC error detection and correction is that it effectively randomizes the burst errors. Due to interleaving, at most one erroneous byte will occur in any given C_2 codeword even when a burst error of one hundred bytes is present. The achievement of full error correction capability of the CIRC system requires, however, the cooperation between C_1 and C_2 decoders through the erasure flags. This cooperation is called enhanced decoding [3,48] and is currently implemented in high-performance CD-ROM drives. A common terminology used to characterize a CIRC decoder is $(t = 2, e = 4)$ which designates at most two symbols corrected at the C_1 level and at most four symbols corrected (by means of the erasure-position method) at the C_2 level, respectively. Ultimately, the success of the correction is also dependent on the implemented decoding algorithm [2,3,77,110].

The maximum allowed BLER according to the CD-ROM standards [53, 80] is 220 counts per second, which means that 3 % of the incoming demodulated frames may contain an error. A $(t = 2, e = 4)$ CIRC decoder is able to correct a maximum burst error of 15 frames [3,77] which is equivalent to 2.65 mm of track length. The ability of a CD-ROM drive to deliver reliable computer data during continuous playback and under adverse playing conditions (e.g. scratched discs) is called playability. Originally defined for an

---

[18]The symbols remained uncorrected at the output of the C_2 decoder can cause undesired noise when an audio disc is played back. As this decoder marks any uncorrected codeword by attaching flags to all corresponding symbols, it will be possible to replace an unreliable 16-bit audio sample by a new one obtained according to a given algorithm. This process is called concealment and consists of either holding a previous reliable audio sample or interpolating linearly between two other reliable audio samples neighboring the one in error [12,45,77]. In order for the concealment to be successful, the scrambling and 2-symbol delay operations separate the adjacent audio samples during encoding.

audio player [3], the CD-ROM playability can be measured either after the CIRC decoder or at the host interface level.

4.8 Turntable motor control

Along with the introduction of the first CD-ROM drives, the use of constant linear velocity (CLV) for playing back the disc has also been accepted. This operating mode of the turntable motor originates in the uniform distribution of data along the disc spiral, which is standardized, and the necessity to maintain a perfectly constant data rate while playing back audio.

However, the performance of the CD-ROM drives has improved over the years and the higher data rates, given by the so-called X-factor, stimulated the use of other techniques for driving the turntable motor. In particular, the CLV operating mode has been replaced by quasi-CLV and further, by pure constant angular velocity (CAV) control. Digital implementations of the corresponding regulators are mostly used in current CD-ROM drives but, for simplicity, only the continuous-time controllers are presented throughout this section. The CLV, quasi-CLV, and CAV controllers are separately discussed. However, a CD-ROM drive may feature all three operating modes, the use of one of them depending on the specific playback situation. For instance, CAV control can be used at very high rotational speeds, quasi-CLV may be switched on during a speed-down action, when a heavily unbalanced disc is detected (see Section 3.3.3), and CLV is usually employed only for audio read-out. Other CD-ROM drives employ quasi-CLV or CAV techniques to spin the disc according to a certain subcode-dependent speed profile. They are designated as adaptive-speed drives and will be discussed at the end of this section.

4.8.1 CLV motor control

Originally intended for audio playback, the CLV operation mode features a very constant data throughput. Although a host system does not require constant data rates form a CD-ROM drive, there were no other solutions when these computer peripherals started to penetrate the market. At that time, an audio player was slightly modified and a data path was added to interface the drive with a host system.

The CLV control remains, however, very important especially because the disc itself, as standardized information carrier, is recorded using a constant

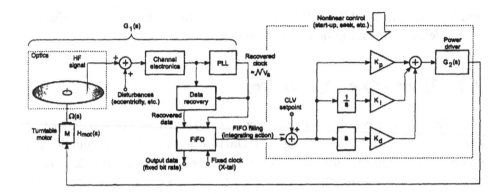

Figure 4.11   Block diagram of CLV turntable motor control.

channel clock. A block diagram of the CLV control is depicted in Fig. 4.11 for a disc spinning at the constant linear velocity $\mathcal{N}v_a$. The recovered clock or a signal synchronous with the PLL voltage-controlled oscillator (VCO) represents a measure of the data rate. Due to linear distribution of data along the disc spiral, the recovered clock frequency is proportional to the disc linear velocity but also to the disc angular speed $\omega_d = \mathcal{N}v_a/r$, at any particular read-out radius r. By using a first-in-first-out (FIFO) buffer to provide output data at a constant clock rate, small fluctuations of the regulated linear velocity can be compensated. In addition, speed variations introduced by the periodical eccentricity[19] of the disc spiral can also be handled if there is sufficient FIFO memory. The FIFO buffer was originally intended for audio playback where a fixed and very stable output data rate[20] is necessary to reconstruct the analog signal.

The amount of data stored in the FIFO buffer at a given moment represents a measure of the angular position $\vartheta = \omega_d t$ of the rotating disc [54,70]. It follows that an ideal integrating action is performed by the FIFO buffer. The phase error information is obtained by continuously examining the degree to which the buffer if filled up. In general, a setpoint of 50 % for the FIFO filling is used in CLV CD-ROM drives.

The phase error is further processed by a compensation network. A PID regulator, as already discussed in Section 3.4.3, is commonly used. The

---

[19] As already discussed in Section 3.3.3, the total spiral eccentricity is the sum of track radial run-out and mechanical eccentricities.

[20] The output bit rate should be an integer multiple of the audio sampling frequency $F_s = 44.1$ kHz. Further details are given in Section 5.4.

transfer function of the CLV open loop is

$$H(s) = rG_1(s)\left(K_p + \frac{K_i}{s} + sK_d\right)G_2(s)H_{mot}(s) \qquad (4.4)$$

where r is the current read-out radius and $G_1(s)$, $G_2(s)$ and $H_{mot}(s)$ are the transfer functions of the decoder front-end, motor driver, and turntable motor, respectively. In general, $G_1(s)$ may be replaced by a real gain because the channel electronics is much faster than the whole motor loop. The motor is of a DC type [62,98] and therefore

$$H_{mot}(s) = \frac{K_t}{(sL_a + R_a)(sJ_{rot} + D) + K_eK_t} \qquad (4.5)$$

when considered between the input voltage and output angular frequency. In case of current-controlled motors, the above relation needs being modified accordingly [105]. The parameters K_t, L_a, and R_a designate the torque constant, armature inductance and armature resistance, respectively. The numerical value of K_t is equal to the back-emf constant K_e when both are expressed in the SI system of units. The total moment of inertia at the motor shaft is denoted by J_{rot} while D represents the damping (viscous friction) coefficient. Typical values for these parameters are $K_m = 0.008\ldots0.014$ Nm/A, $L_a \approx 2$ mH, $R_a = 4\ldots10$ Ω, and a damping constant $D \approx 4 \cdot 10^{-6}$ Nm/s^{-1}. The moment of inertia J_{rot} has numerical values between $5 \cdot 10^{-6}$ and $45 \cdot 10^{-6}$ kg $\cdot$ m^2, depending on the outer diameter (8 or 12 cm) of the disc and on various tolerances of both the disc physical dimensions and material properties.

The Bode plots of a typical PID-based CLV open loop are given in Fig. 4.12 for a 1X drive. The motor rotational frequencies f_{rot} lie within a range determined by the starting diameter D_i and maximum diameter D_o of the program area. These parameters are related through the equation

$$\frac{\mathcal{N}v_a}{\pi D_o} \leq f_{rot} \leq \frac{\mathcal{N}v_a}{\pi D_i} \qquad (4.6)$$

with $v_a = 1.2\ldots1.4$ m/s being the linear velocity of the recorded data and $\mathcal{N}$ being the read-out overspeed factor.

A CLV control loop should basically satisfy four requirements: (i) the FIFO buffer must never become full neither empty during continuous playback, (ii) the constant linear velocity $\mathcal{N}v_a$ must be relatively fast regulated to its nominal value after a seek action (when the disc radius undergoes large variations), (iii) dynamic stability, and (iv) robustness with respect to

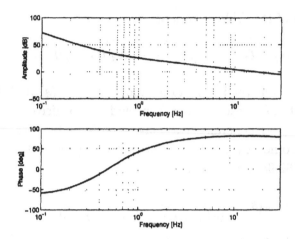

Figure 4.12   Typical Bode plots of the open-loop turntable motor
control, in a 1X CLV CD-ROM drive.

system tolerances. It is important to notice that, apart from eccentricity, the disturbance level is very low and in this respect there is no specific requirement.[21] The loop gain is determined by all conditions mentioned above and depends also on the FIFO length, while the PID gains and corner frequencies should provide the system stability. In general, the bandwidth of the CLV loop [54] is about twice the maximum disc rotational frequency which results from Equation (4.6).

The robustness of the loop in relation to system tolerances is guaranteed by the FIFO buffer, provided that there is no signal saturation along the control path. If saturation occurs, it leads very fast to either an empty or a full data buffer, both situations meaning unreliable output data. Signal saturation during normal tracking can be due to, for example, spinning of a too heavy disc (i.e., with a too large J_{rot}) and it can be avoided by properly choosing the FIFO length and the loop bandwidth.

Another remark concerns the measurement noise which has not been shown at all in Fig. 4.11. This noise, in the form of jitter of the regenerated clock, is introduced at very high frequencies by the PLL and its associated circuits and is completely rejected by the low-bandwidth CLV loop.

---

[21]A specific design case is represented by portable CD-ROM drives which may be affected by gyroscopic effects. These effects introduce additional torques in the motor equation and must be taken into account when designing the control loop of the turntable motor.

Under normal playback conditions (continuous read-out), the basic engine microcontroller does not interfere with the motor loop. Under certain circumstances, however, the microcontroller may help the loop work outside its linear operating region. For instance, the FIFO filling information is not reliable at all during an access procedure because there is no input data while crossings the tracks. In such cases, the motor controller must be reconfigured, e.g. by switching off the FIFO branch. In other cases, the CLV controller might be completely disabled, as it happens during motor start-up when a fixed voltage is usually applied on the motor coils.

4.8.2 Quasi-CLV motor control

As computer data may be forwarded to the host system at a variable bit rate, the CD-ROM drive will not necessarily be of a CLV type. The data discs can be played back in the so-called quasi-CLV mode [103] while for audio discs the original CLV control may still be used.[22] Many current drives are therefore combining both control strategies and, depending on the type of disc being played back, either CLV or quasi-CLV is selected.

A block diagram of the quasi-CLV turntable motor loop is depicted in Fig. 4.13. The data flow is not regulated anymore by a FIFO memory and no phase information is used to control the turntable motor either. The loop controller may have a proportional as well as an integrating branch. However, because we deal here with a velocity loop, a solely proportional regulator is sufficient to guarantee the dynamic stability. It will be also shown later in this section that a strong integrating action is basically not desired, because the loop must feature very short settling times. As the amount of disturbance affecting the controlled quasi-CLV speed is insignificant[23] (with the exception of eccentricity), the integrating branch will neither be needed for disturbance rejection. Nevertheless, an integrating action may be used to reduce the steady-state error of the control loop.

A CD-ROM drive with quasi-CLV control behaves during continuous playback exactly as a CLV drive, i.e., the data throughput is kept constant. However, when seeking across the disc radius, data will become available as soon as the target track and the desired position along the track have been

---

[22]Some CD-ROM drives can buffer audio data that is read out at higher disc speeds and deliver it at 1X to the sound device of the host system (see further Section 6.3). In these cases, audio playback in CLV mode becomes obsolete.

[23]As mentioned in Section 4.8.1, portable CD-ROM drives form a particular design case.

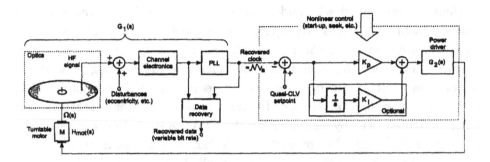

Figure 4.13   Block diagram of quasi-CLV turntable motor control.

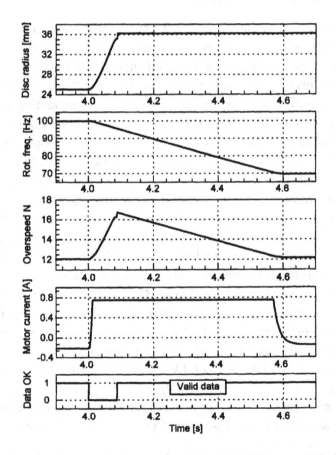

Figure 4.14   Simulated behavior of a quasi-CLV system during an outward-oriented third-stroke seek starting from the inner radius of the disc.

found. This feature is depicted in Fig. 4.14 using a simulation model. Apart from the very last signal, the other four display the same time variations as in a pure CLV drive. The data validity flag, on the other hand, will look differently in a CLV drive where it returns to high logical level only if the overspeed $\mathcal{N}$ has reached again its nominal value (i.e., at about 4.6 seconds in Fig. 4.14). Clearly, the quasi-CLV system can improve the total time needed to access the desired data.

One of the main design criteria of a quasi-CLV loop should be its ability to brake or accelerate the turntable motor very fast during an outward- or inward-oriented seek, respectively. This condition is equivalent to a very short settling time [32,62] and is *only* necessary to avoid exceeding the maximum/minimum decoding speed of the channel electronics. Significantly to remark that fast regulating the velocity towards the desired $\mathcal{N}v_a$ value has minor importance in a quasi-CLV CD-ROM drive (provided that the maximum/minimum decoding speed requirement is fulfilled). The reason behind this affirmation is the capability of a quasi-CLV system to deliver data at variable bit rate.

The condition of short settling time, as stated above, can be translated into a large open-loop gain. Although the tracking (steady-state) error can be reduced, a high proportional gain may produce oscillations in the system. A trade-off between various performance specifications should therefore be considered when designing the quasi-CLV loop. In general, this loop operates with a bandwidth of several Hz.

The quasi-CLV mode (also called pseudo-CLV, variable playback system, etc.) practically dominates the control of the turntable motor in high-end drives. In addition to reducing the access time of the drive, the following two advantages of quasi-CLV mode should also be considered: (i) it allows using of less powerful turntable motors when compared with CLV control at the same overspeed, because the acceleration or braking possibilities of the motor hardly affect the average access performance; (ii) during sustained read-out, features the same (high) bit rate as a classical CLV system of the same overspeed. More information about quasi-CLV systems can be found in [56,101,102,103].

4.8.3 CAV motor control

Once the data rate of the drives has reached very high values, like in 24X or 32X systems, the CAV control has replaced its quasi-CLV predecessor.

The main reason is given by the power dissipated in the motor loop (namely in the turntable motor itself and in the motor driver) which decreases substantially when f_{rot} is kept constant.

The CAV control is practically independent on the clock recovered by the PLL and hence it is independent on the read-out overspeed $\mathcal{N}$. However, the disc rotational frequency should be chosen such that the maximum decoding speed $\mathcal{N}_{max}$ of the electronics is not exceeded. This condition can be written as

$$\frac{\pi f_{rot} D_o}{v_a} \leq \mathcal{N}_{max} \tag{4.7}$$

and thus, the full range of standardized linear velocities of the recorded data $(1.2 \ldots 1.4 \text{ m/s})$ must be taken into account. The disc rotational frequency can either be fixed to a safe value given by $v_a = 1.2$ m/s when substituted in Equation (4.7) or it can be adaptively set, depending on the numerical value of v_a determined during a calibration procedure.[24] Significantly, both CLV and quasi-CLV systems do not have to care about the effective value of v_a because these motor controllers regulate toward the setpoint $\mathcal{N} v_a$.

The control diagram of a CAV loop is depicted in Fig. 4.15. The speed detector, the measurement process itself, and a low-pass filter (LPF) are denoted together by the transfer function $G_1(s)$. The LPF function is needed to suppress the high-frequency components of the measurement noise.[25] When compared to the CLV and quasi-CLV loops, this noise is now playing a more important role because the speed information is obtained at the disc rotational frequency, i.e., within the CAV loop bandwidth.

Also different from the CLV and quasi-CLV loops is the disturbance level affecting the regulated angular frequency $\Omega(s)$. Because no disc parameters, like eccentricity, determine the steady-state rotation frequency of the motor nor any external force or torque needs being opposed (contactless disc read-out), the CAV loop does not require disturbance rejection.[26] The disc mass, which does have a major contribution to J_{rot}, will only influence the start-

---

[24] A calibration procedure is initiated after power-up by the basic engine microcontroller (see also Section 5.5) in order to determine some system parameters, like track pitch, speed v_a of recorded data, etc.

[25] Usually, the speed measurement relies on Hall sensors which are symmetrically mounted inside the turntable motor. If incorrectly positioned, components of high frequency as well as a DC offset are introduced in the feedback signal. A Hall sensor is a device able to generate an electrical current when subjected to a magnetic field [31].

[26] With the exception of portable CD-ROM drives.

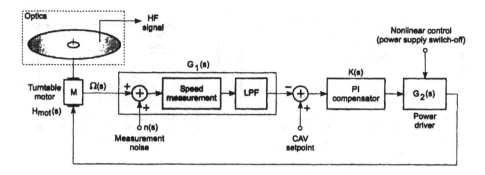

Figure 4.15   CAV turntable motor control loop.

up and stop behavior of the motor because otherwise f_{rot} remains constant during operation.

From the above discussions it follows that a small loop bandwidth (several Hz) is needed for noise rejection in CAV mode and a proportional controller can guarantee the dynamic stability of the loop. For better steady-state error performance as well as for disturbance rejection in portable drives, an additional integrating branch can be added to the proportional regulator.

It is important to remark that both CAV and quasi-CLV systems offer the same performance in terms data access[27] when measured during continuous seeking [101,105]. However, the average data rate of a CAV drive is relatively low, because the overspeed factor $\mathcal{N} = 2\pi f_{rot}r/v_a$ exhibits large variations, given by the read-out radius $D_i/2 \leq r \leq D_o/2$) during continuous playback.

4.8.4 Adaptive-speed CD-ROM drives

Based initially on a quasi-CLV loop, the adaptive-speed control of the turntable motor is intended to further boost the bit rate of the read-out data. Alternative implementations using a CAV loop are also possible [105]. The development of the adaptive-speed CD-ROM drives has been driven by the continuous trend to improve the *average sustained* data throughput measured at the host interface level. Apart from this trend, two system aspects have also played a decisive role.

First, neither CLV nor CAV represents an optimal control strategy when considering both the data rate and the data access performance (see also

---

[27]The performance of CD-ROM drives will be further discussed in Section 7.4.

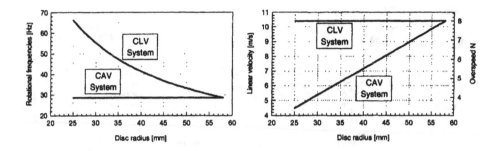

Figure 4.16   CLV and CAV variations of the overspeed and
disc rotational frequency in a CD-ROM system
limited by $f_{rot} = 66$ Hz.

Section 7.4). While the former performance parameter requires CLV playback, the latter will not benefit at all from this control strategy of the
turntable motor. On the contrary, a CAV system features low average data
rate but the data access is extremely fast.

Second, there has been always a limiting factor in the system: either the
channel electronics or the radial and focus servomechanics [101,103]. The
maximum overspeed in the system is determined by the maximum frequency
of the signal which can still be safely processed. Not only the digital channel
electronics contributes to this upper boundary, but the photodetector[28] and
the analog preamplifiers[29] have also a very strong influence. On the other
hand, the drive servomechanics limits the disc rotational frequencies such
that most of the unbalanced discs can remain readable and disc vertical and
radial deviations can still be followed.

An adaptive-speed profile represents an overspeed curve which is adjusted
in real time by the drive microcontroller during data playback. Graphically,
such a curve may be situated anywhere between the CLV and CAV speed
characteristics. As an example, these characteristics are plotted in Fig. 4.16
for a CD-ROM system limited by $f_{rot} = 66$ Hz. The region between the
CLV and CAV plots can be used for control purposes.

The various speed setpoints which determine an adaptive-speed curve are
dependent on the read-out subcode timing[30] and are intended to be followed

---

[28]See Section 5.2.

[29]See Section 5.3.

[30]For instance, a new speed setpoint may be provided for each 2-minute increment of
the read subcode timing. Details about the construction of the subcode timing are given
in Section 6.1.

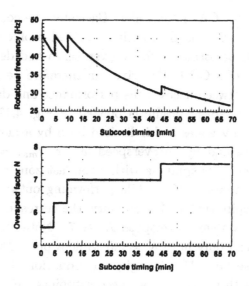

Figure 4.17 Zoned adaptive-speed profile for a CD-ROM system limited by the disc rotational frequency at 46 Hz and by the decoding electronics at $\mathcal{N} = 8$.

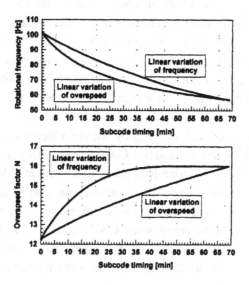

Figure 4.18 Continuous adaptive-speed profiles for a CD-ROM system limited by the disc rotational frequency at 100 Hz and by the decoding electronics at $N = 16$. Both profiles depend linearly on the disc radius but are plotted versus subcode timing.

by the quasi-CLV (or CAV) controller. Basically, the common attribute of all adaptive-speed drives is that their overspeed profile can be optimized to obtain the best performance for a given servo and decoder electronics. The specifications of a CD-ROM drive can therefore be tuned towards an optimal trade-off between access time performance and data rate.

A zoned adaptive-speed profile is plotted in Fig. 4.17. The system is based on 4 quasi-CLV zones and is limited both by a maximum rotational frequency of 46 Hz and by an overspeed factor $\mathcal{N}_{max} = 8$. Significantly to notice that the disc is spinning within the last zone at $\mathcal{N} = 7.5$, which provides sufficient margin [103] while performing outward-oriented third-stroke seeks[31] Despite the 46-Hz boundary, the corresponding drive is still able to reach an average[32] overspeed $\mathcal{N} \approx 7$. Note that a CLV system limited at 46 Hz can only have $\mathcal{N} \approx 5.5$ while a pure CAV drive performs even worse, because $3.5 \leq \mathcal{N} \leq 8$ as given by Equation (4.7). A zoned profile can be obtained with relatively few speed setpoints which might represent an advantage in case of analog quasi-CLV implementations.

Other two examples of adaptive-speed profiles are given in Fig. 4.18. They employ either an overspeed factor $\mathcal{N}(r)$ or a disc rotational frequency $f_{rot}(r)$, both depending linearly on the radius of the read-out track. When plotted versus subcode timing, these dependencies display a second-order curvature. As the data rate is proportional with $\mathcal{N}$ (see also Sections 6.2 and 7.4.1), the linear variation of rotational frequency represents a better choice for higher data throughputs. On the other hand, the linear variation of the overspeed factor might be easier to implement in a quasi-CLV than in a CAV loop.

Other adaptive-speed drives employ straightforward combinations between CAV and CLV control (partial-CAV drives). Depending on the read-out subcode timing, the angular frequency of the turntable motor is chosen to accommodate the servo capabilities and the speed range of the channel electronics. There are also drives that combine an adaptive-speed profile during continuous data playback with CAV motor control during continuous data access [105]. These drives can separately be optimized for access performance, data throughput and, very important, power consumption.

---

[31]Due to the linear distribution of data along the disc spiral, the overspeed factor reaches a peak value at the end of the sledge-actuator displacement (see also Fig. 7.1). For good access performance, these peaks must be situated within the decoding speed range of the channel and data path electronics.

[32]See further Section 7.4.2 for details regarding the calculation of the average bit rate.

Chapter 5

Other basic-engine building blocks

We have only discussed until now the optics, servo-mechanics and the channel decoder as three essential parts of a CD-ROM basic engine. However, data delivery would not be possible without generating a laser beam, converting the optical power into electrical signals and without the supervision of a central microcontroller. In addition, any CD-ROM drive should also play back audio discs, which calls for an adequate digital-to-analog (D/A) interface.

5.1   Semiconductor lasers

The laser beam used in CD-ROM systems is generated by an aluminum-gallium-arsenide (AlGaAs) semiconductor device [10,119]. The emission of light within a laser is due to the transition of atoms between two adjacent energy bands [1]. The wavelength of the emitted radiation obeys the relation $\lambda = \hbar/\Delta E$ where ΔE is the energy difference between the two transition levels and $\hbar = 6.6256 \cdot 10^{-34}$ Js is the Planck constant. The laser, as an optical oscillator, uses an amplifying medium placed between two mirrors (which form a resonator) and an energy source to stimulate the emission of

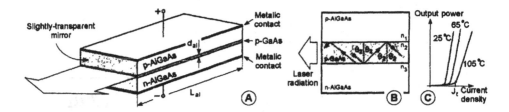

Figure 5.1   Schematic construction of a semiconductor laser (A), longitu-
dinal cross-section through the device (B), and output power
as function of the current density (C).

radiation [76,123]. The source pumps energy into the medium, moving the
atoms into a higher energy level. From this excited state, the atoms fall
spontaneously to their ground state while emitting a photon.

The laser process depends on the following conditions: (i) the photon
stimulated emission should lead to light amplification, which is ensured by
carefully choosing the resonator length to allow the formation of standing
waves; (ii) a population inversion should be created between two appropriate energy levels – achieved by the energy pump; (iii) seed photons of
proper energy and direction are needed to initiate the emission – realized
by spontaneous transitions; and (iv) coupling a fraction of the laser light
out of the oscillator, which is achieved by allowing a slight transparency for
one of the resonator mirrors.

Several unique features [1] characterize the laser beams: (i) the radiation
is intense, coherent[1] and monochromatic; (ii) the beams are highly direc-
tional and collimated, experiencing very small angular dispersion during
propagation; and (iii) the beams can be focused very sharply.

A laser diode, as used in optical recordings, is based on the stimulated
emission of photons which takes place in the neighborhood of the junction
between two semiconductor materials. Several semiconductor structures
can be used [109]. The laser employed in CD-ROM systems is of a double-
heterostructure (DH) type and has a wavelength $\lambda = 780\,nm$. A schematic
construction of this laser is presented in Fig. 5.1-A. The stimulated emission takes place in the thin p-type layer of GaAs (called the active layer)

---

[1]The coherence (temporal and spatial) represents a measure of the phase uniformity
across the optical wavefront. In the process of stimulated emission, each photon added to
the stimulated radiation has a phase, polarization, energy and direction identical to that
of the amplified light wave in the laser cavity.

sandwiched between two p- and respectively n-type layers of AlGaAs. Under forward DC bias[2], electrons are injected from the n-side and holes are injected from the p-side into the transition region. As a result, the GaAs region contains a large concentration of electrons in the conduction band and a large concentration of holes in the valence band, which is the condition for population inversion. The stimulated laser radiation is initiated when the current density in the p-layer exceeds the threshold value [109]

$$J_t = \frac{J_0 d_{al}}{\eta} + J_0 \frac{d_{al}}{g_0 \eta \Gamma} \left[\alpha + \frac{1}{L_{al}} \ln \left(\frac{1}{R_{mirr}} \right) \right] \qquad (5.1)$$

where J_0 and g_0 represent the nominal current density and nominal optical gain[3], respectively, d_{al} is the thickness of the active layer, η is the quantum efficiency[4], Γ is the confinement factor[5], α represents the loss in optical energy per unit length (absorption coefficient), L_{al} is the length of the active layer, and R_{mirr} stands for the reflectance of the resonator mirrors. In addition to Equation (5.1), the reflection angle θ_2 inside the resonator [59] should satisfy the condition $\sin \theta_2 \geq \max(n_1, n_3)/n_2$ with n_1, n_2, n_3 being the refractive indices as shown in Fig. 5.1-B. When fulfilling this condition, the laser beams are trapped by internal reflections within the active layer. Other characteristics related to the output power of the laser diode and its emission spectra can be found in literature [8,109,123].

As a final remark, we mention that a laser diode slowly deteriorates in time while continuously under a DC bias. The consequent decrease of the light intensity can be avoided by capturing a fraction of the emitted light on a photodetector and use this signal in a feedback loop to increase the electrical current through the device. Such a laser controller is used in all CD-ROM drives.

5.2 The photodetector

The laser beam reflected by the disc surface is directed towards photodetectors (see Section 2.2). These are semiconductor devices that can transform

---

[2]This operation mode is usually called continuous wave (CW), as opposed to the operation under a frequency-modulated current superposed on the DC bias.

[3]The gain g is a measure of the incremental optical energy flux per unit length.

[4]The quantum efficiency of a photonic emitting device is given by the number of photons generated per electron-hole pair.

[5]The confinement factor is the ratio of the light intensity within the active layer to the sum of light intensities both within and outside the active layer.

incident optical power into electrical signals. A CD-ROM system uses an arrangement of several photodiodes to generate electrical currents which will be further combined by dedicated electronics into focus/radial error and HF signal, respectively. The two common photodiode arrangements have already been shown in Fig. 2.11, for the twin-spot radial tracking combined with astigmatic and single-Foucault focus detection, respectively.

The operation of a photodiode, and of a photodetector in general [109], involves three steps: (i) carrier generation by incident light; (ii) carrier transport and/or multiplication by a current gain mechanism; and (iii) interaction of the generated current with the external circuit to provide the output signal. The performance requirements are given in terms of high sensitivity, low noise, wide bandwidth, high reliability, and low cost [8].

When a photodiode is illuminated, the incident photons are absorbed to create electron-hole pairs as shown in Fig. 5.2-A and -B. The diode is operated under reverse-biased conditions, when the depletion region[6] is quite wide and a small reverse saturation current (also called dark current) flows through the device. When the p-n junction is excited by the incident light, photons are absorbed mainly in the depletion layer but also in the neutral regions, particularly on the top where the light hits the device. The carriers (electron-hole pairs) generated in the depletion region are accelerated in opposite directions by the reverse bias and give rise to a photocurrent. The magnitude of this current (with contributions from both the depletion layer and the n^+-region, called bulk) depends on the external quantum efficiency[7]

$$\eta_{ext} = \eta_{int}(1 - R_{coat})\left(1 - \frac{e^{-\alpha W}}{1 + \alpha\sqrt{D_p\tau_p}}\right) \tag{5.2}$$

where R_{coat} is the reflection coefficient of the incident surface (usually a dielectric coating), α is the absorption coefficient of the bulk semiconductor material, W is the width of the depletion layer, D_p is the diffusion coefficient

---

[6]The depletion region (also called the space charge region) is formed around the junction between two semiconductors differently doped and is characterized by zero-densities of the mobile carriers (electrons and holes).

[7]The external quantum efficiency of a photodetector is given by the number of carriers (electron-hole pairs) collected to produce the photocurrent, divided by the number of incident photons. The internal quantum efficiency η_{int} is the number of created electron-hole pairs divided by the numbers of absorbed photons and is usually very high, if not unity, in defect-free materials [8].

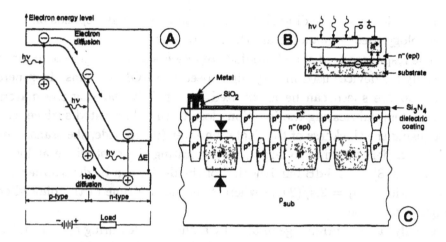

Figure 5.2   (A) Energy bands and electron-hole pair generation in a photodiode under reversed bias. (B) Cross-section through a junction photodiode. (C) Cross-sectional view of a state-of-the-art photodetector.

for holes, τ_p is the lifetime of the excess carriers[8], and η_{int} denotes the internal quantum efficiency.

Practically, not all parameters involved in Equation (5.2) can be trimmed to increase the photodiode efficiency. For example, the absorption coefficient α, which is a property of the semiconductor material, cannot be changed anymore once the bulk semiconductor (the n^+-region in Fig. 5.2-B and -C) has been chosen. A similar remark holds for the product $D_p\tau_p$. It follows that η_{ext} can basically be improved by choosing an adequate coating with $R_{coat} \cong 0$ and by enlarging the depletion layer. The thickness W of the depletion layer can be increased by sandwiching an intrinsic[9] layer between the two p- and respectively n-type semiconductor films already depicted in Fig. 5.2-B. This construction is called *pin*-diode and, when compared to a classical *pn*-junction, it can be optimized for both quantum efficiency and frequency response. Common values [8] for W are between $1/\alpha$ and $2/\alpha$ with $\alpha = 10\ldots10^6\,\mathrm{cm}^{-1}$ depending on the semiconductor material but also on the incident wavelength λ. However, the desired thickness of the

---

[8]Typical values [8] for a semiconductor material are $D_p = 13\ \mathrm{cm}^2/\mathrm{s}$ and $\tau_p = 10^{-7}$ s.

[9]An intrinsic semiconductor (i.e., not doped) contains, per unit volume, the same number of electrons in the conduction band as number of holes in the valence band.

depletion layer may be difficult to obtain, as W strongly depends on the IC technology used to manufacture the photodiode.

The response speed of a photodiode represents another figure of merit, especially for devices used in very high-speed CD-ROM systems. In general, the response speed can be maximized by carefully choosing the material parameters, fabricating the junction close to the illuminated surface, compromising the thickness of the depletion layer (if too wide, the transit time of the carriers becomes to large), and operating the photodiode at low reverse bias [8]. The 3-dB bandwidth[10] of the device can be determined with the relation $f_{3dB} = 2.4/(2\pi t_r)$ where t_r is the rise time of the photodiode impulse response.

Finally, as the currents generated by a photodiode during operation have small amplitudes ($1\dots20$ μA), the associated noise should also be taken into account. It can be shown [8,109] that the resistance or Johnson noise dominates in these devices[11] and it can be minimized by optimizing the circuit parameters. Under these circumstances, the low-frequency measurement noise from the position control loop of Fig. 3.4 may be neglected.

A cross-section through a state-of-the-art photodetector (i.e., combination of photodiodes) for high-speed CD-ROM systems [35] is represented in Fig. 5.2-C. The depletion region of each diode is formed between the p-type regions surrounding each of the n-type buried diffusion regions. The device features a common anode for all diodes through the p-type surface region.

5.3 Preprocessing of the photodetector signals

The preprocessing of the generated photocurrents is performed by analog electronics which is usually incorporated in one integrated circuit (IC), possible together with the photodetector itself.

There are several reasons for which an interface is needed between the photodetector and the servo and decoding subsystems, respectively.

---

[10]The 3-dB bandwidth is defined as the frequency at which the photocurrent amplitude drops to half of its value at low frequencies.

[11]The Johnson noise results from the random motion of carriers that contributes to the dark current of the device. Other disturbance sources are the generation-recombination process which gives rise to shot noise, the surface and interface defects in the bulk semiconductor (determining the flicker or $1/f$ noise), and the optical signal itself, which generates the quantum noise.

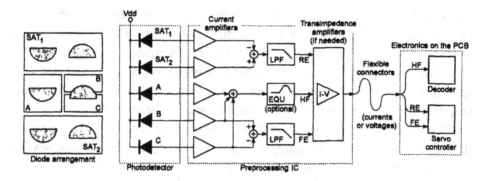

Figure 5.3 Preprocessing functions in a CD-ROM system based on 3-beam single-Foucault optics.

First of all, a current will only be generated by a photodiode if the latter is connected in a closed circuit. The interface acts therefore as load for the reverse-biased diode.

Second, the generated currents are of quite small amplitudes, typically of $1 \ldots 20\,\mu A$. The servo and decoder functions are normally implemented on the main printed circuit board (PCB) whereas the photocurrents are generated at a reasonable distance, on the moving optical head. As flexible connectors are used to carry the diode signals to the PCB, a good signal-to-noise ratio[12] (SNR) can only be obtained if the generated photocurrents are sufficiently amplified by an interface circuit.

Finally, the output of the photodetector consists of independent currents which may not represent always the right choice for the input of a servo or decoder IC. As already discussed in Sections 2.5 and 2.6, these currents need being cumulated or subtracted to provide the focus and radial servo signals as well as the HF signal. An additional function of the interface circuitry is to filter out the high-frequency components from those currents which are used for servo purposes. Also, some preprocessing ICs incorporate a channel equalizer to operate upon the HF signal [27].

An example of preprocessing functions associated with a 3-beam single-Foucault optics is schematically depicted in Fig. 5.3. The photodiode currents are amplified and further combined into radial/focus error (RE/FE) signals and high-frequency signal, respectively. The resulted currents are

---

[12]We refer here to the signal-to-noise ratio of the photocurrents with respect to all disturbances, like digital clocks, not originating from the photodetector itself.

converted to voltages and sent through flexible connectors to the PCB. Another common CD-ROM architecture does not feature RE/FE subtraction circuits in the preprocessing IC but they are integrated together with the servo functions [27]. Yet another possibility is to send currents and not voltages through the flexible connectors and consequently, the transimpedance amplifiers will not be part of the preprocessing IC anymore.

We shall not end this section without mentioning the extreme care which should be taken when designing the preprocessing circuitry. The several microamperes at the input are very sensitive to PCB and IC layout. The offsets must be eliminated, if possible, or carefully controlled over a relative large range of ambient temperatures (usually $5 \ldots 55\,°C$) and the HF path should be designed to provide a constant delay for all EFM patterns. Last but not least, the parasitic capacitances of the flexible connectors are posing problems, especially in high- and very high-speed CD-ROM drives.

5.4 Audio circuits

Despite its use in a computing environment, a CD-ROM drive must still obey the digital audio standard [50,79]. In fact, from a basic engine standpoint, there is absolutely no difference between data and audio discs. It is only the 24 user bytes from each demodulated frame, as depicted in Fig. 4.8, which contain either computer data or sampled audio signal but the rest of the encoding process, like scrambling, error correction, etc., remains unchanged. The CD-ROM standard [53,80] does only represent a functional upgrade (see further Chapter 6) and the playback of an audio disc is still possible, if not desired, in any CD-ROM drive.

Prior its recording, the audio signal is digitized simultaneously on two stereo channels at a sampling rate $F_s = 44.1$ kHz, which is sufficient for reproductions of maximum frequencies up to 20,000 Hz.[13] Each single-channel audio sample contains 16 bits which leads to a signal-to-noise ratio (SNR) of 98 dB [82]. A good overview of the CD audio recording process is presented in [70] and of the digital audio systems, in general, in [54,83,116].

Following the decoding process already discussed throughout Chapter 4, a digital stream of audio data will become available. This stream contains serial patterns of 16-bit stereo samples which alternate (time-multiplexed)

---

[13]It is generally considered that frequencies between 5 and 20,000 Hz are sufficient for high-fidelity reproduction of audible sounds. When frequencies above 20 kHz are completely cut off, they are in proportion of 99 % inaudible [70].

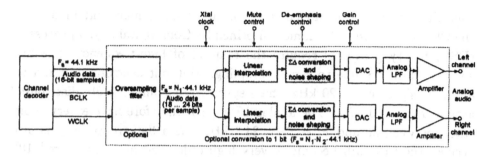

Figure 5.4   Block diagram of audio functions.

for the left and right channel, respectively. The bit rate of this data stream equals $2 \times 16 \times 44.1 \cdot 10^3 = 1.4112$ MHz and an additional clock (BCLK) of this frequency is used to mark the position of each bit. A second signal which accompanies the serial data stream is the word clock (WCLK) which toggles between 0 and 1 for each 2 bytes representing a left/right sample.

A block diagram of the audio circuits is given in Fig. 5.4. Basically, only the white-filled blocks are necessary to convert the incoming serial data into an analog signal. The decoder outputs digital audio data in a standardized[14] IC format [27] and the digital-to-analog converter (DAC) produces an analog-like wave which resembles the original audio signal. Different techniques, like weighted resistors, R-2R ladders, integration or current matching are used to implement the DACs [70,82,83]. The final touch is given by the analog low-pass filter (LPF) which removes the frequency components situated above 20 kHz. These components are both due to fast commutation inside the DAC (switches, comparators, etc.) and to symmetrically mirroring of the audio frequency spectrum around the sampling frequency [34,74,86]. Additional functions like muting, de-emphasis[15] or gain control are also provided.

The oversampling filter form Fig. 5.4 is marked as optional because its bypassing does not prevent the analog audio to be retrieved. However,

---

[14]The Philips format, for example, is called Inter-IC Sound (I²S) and assigns only one clock (BCLK) period for each audio bit. The serial pattern is coded in 2's complement and begins always with the most significant bit (MSB). It is also allowed to reverse the polarity of the word clock (WCLK), which means that a zero logic of WCLK does not necessarily mark a left-channel sample.

[15]Because the high-frequency contents of music is relatively small, a first-order pre-emphasis (+20 dB/decade) in this frequency range may be applied during recording. The reverse operation (de-emphasis) needs being performed during playback [50,79].

this filter is present in all current CD audio equipment and in most of the CD-ROM drives. Fundamental principles from digital signal processing (DSP) theory [34,74,86] require the removal of high-frequency (aliasing) spectra which are due to signal sampling. Without oversampling and for an audio signal up to 20 kHz, these spectra are symmetrically distributed around $F_s = 44.1$ kHz. The analog LPF must therefore have a very sharp cutoff between 20 and $F_s - 20 = 24.1$ kHz. As a flat group delay is also needed, the LPF will become of very high order and quite complex [118]. The design requirements can be relaxed if an oversampling filter is used to shift the aliasing components around $N_1 \cdot F_s$ with N_1 being a power of 2 (usually $N_1 = 4$ or $N_1 = 8$). Another advantage of the oversampling filter is the expansion of its output resolution to $18\ldots24$ bits which leads to the redistribution of the overall quantization noise [82] between 0 and $N_1 F_s/2$.

Finally, the (also optional) conversion to 1 bit from Fig. 5.4 will further increase the sampling rate of the digital audio signal by a factor N_2 but its main role is to further reduce the noise power density in the audio band. This technique, called noise shaping[16] provides a mean to distribute the noise, in a frequency-dependent way, between 0 and $N_1 N_2 F_s/2$. The noise shaper can be used without preceding oversampling filter (i.e., $N_1 = 1$) and the oversampling ratio N_2 can reach very high values, like 128 or even 256.

5.5 The microcontroller

The link between all building blocks within a CD-ROM basic engine is provided by a microcontroller (see Fig. 1.3) running under firmware super-vision. The firmware, which is stored in a non-volatile memory, represents a collection of interconnected logical decisions which are designed to master various activities in the system. A schematic overview of the basic firmware functions is depicted in Fig. 5.5.

Two basic activities are carried out right after the drive has been powered up: the initialization coefficients are sent to all programmable components,

---

[16]The noise shaper, also called single-bit DAC, bit-stream DAC or Sigma-Delta ($\Sigma\Delta$) modulator, is a 1-bit quantization device whose output data is sampled at very high frequencies when compared to the input sampling frequency. The difference between the output and input signals is fed back via a specific transfer function which creates a pulse density modulation (PDM) output stream. A benefit of the device at low (audio) frequencies is its excellent noise suppression. In fact, the total noise is shaped in frequency and pushed outside the audio band. Further details about $\Sigma\Delta$ modulation can be found in [19,54,71,72].

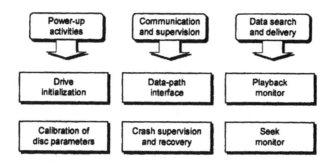

Figure 5.5 Overview of firmware functions.

namely the servo and channel decoder ICs, and a calibration procedure is executed. This procedure determines the disc parameters, like the linear velocity v_a of recorded data, track pitch q, total playback time, and the disc size (12 or 8 cm). A good calculation of the disc parameters is of paramount importance for the success of a seek action. The disc size is needed for adjusting the control loop of the turntable motor.

During normal operation, the two microcontrollers of the basic engine and data path are communicating with each other based on a protocol between their firmware algorithms. The communication takes actually place at this protocol level, even if the basic engine and data path are monitored by a common microcontroller.

The requests issued by the data path are interpreted and converted into basic engine commands. In fact, the data path only asks for a particular information which is located on the disc at a given subcode timing. The basic engine microcontroller determines with Equation (3.15) the physical location of the requested data, assigns necessary tasks to servo and channel decoder ICs, and monitors the seek and the consequent playback actions.

A very important role of the microcontroller is the overall supervision of the state of the basic engine. The complexity of the system, the required accuracy combined with the spread of system tolerances, the disc irregularities, and the environmental conditions can easily lead to various undesired situations. Any function has a certain probability for crash and, for this reason, the firmware monitors continuously all activities and initiates recoveries if necessary. From this point of view, one of the main concerns during a crash is to avoid any disc damage.

Chapter

6

The CD-ROM data path

As already indicated in Fig. 1.3, an interface electronics is provided between the CD-ROM basic engine and the host system. Any disk-based storage device makes use of such an interface, usually referred to as data path. Its role is basically two-fold: to translate the host commands into instructions specific to that particular storage device (drive) and to regulate the data flow between drive, as a peripheral, and the host system.

In any CD-ROM drive, the user information flows to only one direction: towards the host system[1] The type of the built-in data path depends on the interface protocol[2] associated with the host. This protocol does not influence the two data-path functions mentioned above but it does require

---

[1]In CD-R and CD-R/W as well as in all magnetic and magneto-optical (MO) disk drives, the data flow is bidirectional.

[2]Two interface standards are commonly used in personal computers to connect large-capacity disk drives: IDE (Intelligent Drive Electronics) and SCSI (Small Computer System Interface). IDE can only accommodate at most two peripherals (drives) when connected in parallel on the same cable but up to four IDE interfaces (Enhanced-IDE) may be present in one PC. IDE is relatively cheap and easy to use. The SCSI interface supports up to seven drives in parallel and it offers a better performance, including data transfer speed, in real multitasking computing. Some details about both interfaces can be found, for instance, in [91,112].

different hardware and software implementations of the interface itself. The host requests for recorded information are finally converted by the CD-ROM data path into subcode timing, which is a suitable reference to rapidly locate user data on the disc.

An additional function of any CD-ROM data path is to convert the read-out digital information, originally designed for audio recordings, into a format for general data storage. This format, which is not tied to any particular application, has also been standardized [52,80].

Finally, a fourth function of the data path in a CD-ROM drive is to add one more error detection and correction layer. Because the original audio standard does not fulfill the reliability level [110] required by the computing industry (i.e., a bit error rate of at most 10^{-12}), the CD-ROM standard provides additional correction capabilities, usually referred to as third-layer error correction.

6.1 Compact Disc subcode format

As already shown in Fig. 4.8, a 1-byte subcode symbol is heading each user frame after demodulation. The subcode symbols contain various control information which was initially intended for audio playback. For instance, the user can determine which piece of music is being listened to, how long a piece of music is and whether it has been recorded with pre-emphasis or not, etc.

For a CD-ROM system, the subcode information plays also an essential role. First, a subcode block and a CD-ROM data block (see further Section 6.2) are both based on 98 successive demodulated frames. Second, by inserting a special bit pattern in the subcode block, the presence of either CD-ROM or audio data can be detected. And finally, the time reference or subcode timing, which linearly marks the position of each group of 98 user frames along the disc spiral, remains valid for any CD-ROM disc.

The arrangement of the 98 demodulated frames which provides a subcode block is depicted in Fig. 6.1-A and the internal format of this block is given in Fig. 6.1-B. The first two subcode symbols S_0 and S_1 contain synchronization information for the whole block. Because these symbols use two out-of-rule[3] patterns from the EFM conversion table, they must be

---

[3] An out-of-rule EFM pattern satisfies the (2,10)-RLL requirements (see Section 4.6) but it does not represent a valid modulation code for any of the 256 data symbols.

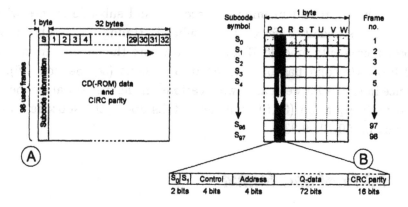

Figure 6.1   Subcode format and Q-channel structure.

detected before demodulation takes place. The following 96 symbols carry bit-wise information from each of the 8 subcode channels which are designated as P, Q, R, S, T, U, V, and W (there is no relation to the P and Q parity from CIRC). The subcode channels are read as shown in Fig. 6.1-B and each of them, except P, contains provision for error detection and/or correction. The Q-channel, for instance, ends with a 16-bit parity word for cyclic redundancy check[4] based on the polynomial $g(x) = x^{16} + x^{12} + x^5 + 1$.

On most audio discs, only the P and Q channels contain information whereas the others are recorded with zeros. The Q channel, which is essential for audio as well as CD-ROM systems, contains information about the position of the associated data block from Fig. 6.1-A along the disc spiral. In addition, a data disc is designated by writing 01x0 into the 4-bit control area. Because the linear velocity v_a of the recorded data remains constant along the entire disc spiral, the position information recorded into the Q-

---

[4]The cyclic redundancy check (CRC) code is an error detection method which adds $n - k$ parity bits to a user word of k bits. The user words are given by a linear code (see Section 4.7) and the resulting n-bit words form a cyclic code. The latter property implies that all codewords are multiples of a generator polynomial $g(x)$ whose degree equals $n-k$. At the receiving side, the codeword is divided by $g(x)$ and a remainder different from zero indicates the occurrence of at least one error. The division can easily be implemented by using a linear-feedback shift register which is, in fact, a recursive digital filter [34,86] whose taps correspond to $g(x)$. This filter can also be used to calculate parity bits before transmitting the codeword. The parity is obtained by feeding the k-bit user words followed by $n - k$ trailing zeros to the input of the filter. The CRC code is capable to detect all burst errors less than or equal to the length $n - k$ of the parity word. Details about CRC codes can be found in [54,66,83].

data can be expressed in minutes, seconds, and subcode frames within a second and is usually referred to as subcode timing. The subcode frame counter increments modulo $7350/98 = 75$ where the channel frame rate of 7350 frames/second has been taken into account (see Section 4.6). The distance in tracks between any two locations on the disc spiral can be determined with Equation (3.15). Further details about the subcode structure are given in [50,70,79,83].

6.2   CD-ROM data format

Obeying the same arrangement from Fig. 6.1-A, a CD-ROM data block is formed by collecting user symbols from 98 consecutive demodulated frames. A CD-ROM block carries thus $98 \times (33 - 9) = 2352$ bytes (9 bytes out of 33 were already reserved for the subcode symbol and CIRC parities).

Each CD-ROM block starts with a 12-byte synchronization pattern[5] and continues with a 4-byte header as depicted in Fig. 6.2. The first three bytes of the header contain the absolute time in minutes, seconds, and block number within a second, indicating the position of the associated data block along the disc spiral. This information is copied from the Q-channel with a maximum allowed offset of ± 1 second, being synchronized with the subcode timing. For this reason, the block number is also designated as frames within a second. The last header byte identifies three possible CD-ROM modes. *Mode 0* indicates an unused area and the remaining $2352 - 16 = 2336$ bytes are filled with zeros. *Mode 2* defines a block of 2336 data bytes and is rarely used in this form because no additional error correction is provided. However, the CD-ROM Extended Architecture[6] (CD-ROM/XA) defines a *Form 1* and a *Form 2* of this mode which do contain provision for checking the data integrity [55,83].

Finally, *Mode 1* defines a data block which is mostly dedicated for computer storage. This format is also represented in Fig. 6.2. There are 2048 bytes of user data while the trailing 288 bytes, also denoted as auxiliary

---

[5]See further Fig. 6.4-A.

[6]The CD-ROM/XA is an extension of the Yellow Book standard (see Fig. 1.1) and defines a new type of data track [81]. An XA track may contain computer data, compressed audio/video data, and/or still pictures. The Red Book CD-DA cannot be placed on an XA track which, when used for compressed audio, delivers a data throughput of 175.2 kB/s. Clearly, special hardware is needed to decode the various types of XA data. CD-i, Video-CD and Photo-CD are all types of CD-ROM/XA and this format has been standardized in the White Book [55].

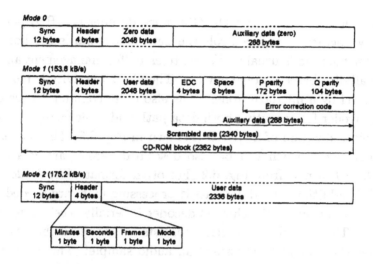

Figure 6.2 Basic formats of CD-ROM data .

data, are reserved for error detection and correction information. The whole block, except the 12-byte sync pattern, is scrambled. Because the computer data is not necessarily randomly distributed (i.e., the same byte may be repeated for a long data sequence), the scrambling facilitates the averaging of the EFM power spectrum during encoding. The *Mode 1* data block contains a 4-byte field for error detection code (EDC) and a 276-byte field for error correction code (ECC). The latter is divided into P and respectively Q parity and operates upon a block of 2340 descrambled bytes (see further Section 6.3). The EDC operates upon the 2064 bytes of sync, header and user data.

The last remark related to the CD-ROM format concerns the user data rate, also called data throughput or, as in digital communications, bit rate. For the most used *Mode 1* data, the speed of the user information received by the host system is 2048 bytes $\times$ 75 s^{-1} = 153.6 kB/s. Usually, a value $B_a = 150$ kB/s is considered in literature. The notation B_a designates the data throughput of a 1X CD-ROM drive (i.e., having a data disc rotating at the audio constant linear velocity v_a).

6.3 Block decoders

Following the extraction of the user frames by the channel decoder, the CD-ROM data is further processed according to its specific format already discussed in Section 6.2. A CD-ROM block decoder (CDBD) performs all necessary operations and controls the data flow towards the host system. The latter function is usually referred to as buffer management and is common to all disk-based storage devices. The CDBD activity is monitored at system level by a microcontroller. It is also quite common to use only one microcontroller for executing both data path and basic engine firmware.

The block diagram of a CDBD is shown in Fig. 6.3. The reference data path operations, which will be also described herein, are related to the basic *Mode 1* format from Fig. 6.2. For other formats, like *Mode 2* of CD-ROM/XA, additional (and dedicated) processing units are needed.

The data output by the channel decoder is serially received by an input interface. The very first operation consists in swapping the order of any two bytes that would have formed an audio sample[7]. This rearrangement is based on a left/right audio rule and is synchronized with the rising and falling edges of the word clock (WCLK). The configuration of a rearranged data block, including the corresponding sync and header bytes, is shown in Fig. 6.4-A. The subsequent operations are performed upon each data block and follow the detection of the corresponding 12-byte synchronization pattern. All 12 sync bytes are discarded after synchronization.

The data arranged as indicated in Fig. 6.4-A is further serially fed to a descrambling circuit which uses a presetable linear-feedback shift register [53,80] based on the polynomial $x^{15} + x + 1$. The descrambling operation takes place as depicted in Fig. 6.4-B and starts with the least significant bit right after the sync pattern. The resulted data is further loaded into a memory. From this point onwards, all subsequent operations are performed at byte level by extracting from memory only those bytes which are needed.

Following descrambling, an error detection is carried out. A cyclic redundancy check[8] (CRC) is performed using the generator polynomial $g(x) = (x^{16} + x^{15} + x^2 + 1)(x^{16} + x^3 + x + 1)$ and the provided 4-byte EDC. If the polynomial associated with the combined sync, header and user data is divisible by $g(x)$, there is a high probability that the received user data will not contain any error and it will be further forwarded to the host system.

---

[7]When audio information is recorded on disc, each data frame contains a sequence of 6 left/right 16-bit audio samples (words). The byte order of any two successive left/right samples, as they are output by the channel decoder, is MSB, LSB, MSB, LSB. In case of data recording, this byte order remains unchanged but it needs being reversed into LSB, MSB, LSB, MSB before any other operation in the block decoder takes place.

[8]See also Section 6.1 for additional CRC information.

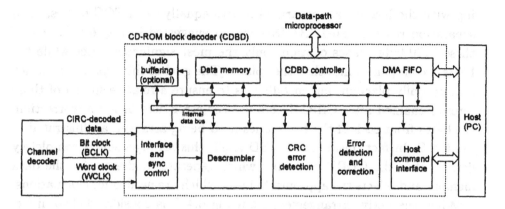

Figure 6.3 Block diagram of a CD-ROM block decoder.

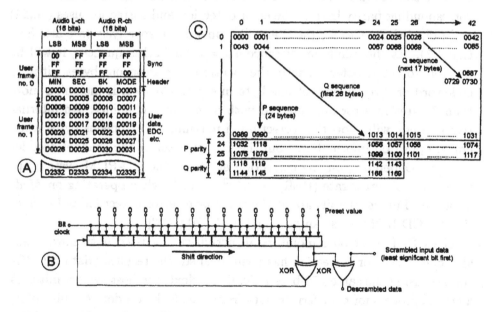

Figure 6.4 The sync, header and user data rearranged by the input interface (A), descrambling linear-feedback shift register (B), and Reed-Solomon correction strategy for CD-ROM data (C).

Significantly to notice that the ECC field of P and Q parity bytes is not subjected to cyclic redundancy check.

If errors are signaled during CRC, a correction mechanism based on the 276-byte ECC is further employed. First, all 2340 descrambled bytes, starting with the header, are separated in two equally-sized ECC planes. This separation relies on an LSB/MSB rule as suggested in Fig. 6.4-A. One of these 1170-byte planes contains only the most significant bytes while the LSBs are all placed into the other plane. The goal of this separation, which took initially place during encoding, is to ensure a relative spread of those errors remained after CIRC decoding. For error detection and correction, the bytes from each of the two planes need being arranged as illustrated in Fig 6.4-C. In fact, inside the CDBD itself, this arrangement is completely fictive. It is the CDBD controller which generates addresses for the data memory and selects the appropriate bytes which are needed for calculations.

When the fictive arrangement from Fig 6.4-C is considered, two different kinds of byte sequences can be formed. Designated as P and Q, each of these sequences forms a Reed-Solomon codeword together with the two corresponding parity bytes. The error detection and correction uses (26,24) and (45,43) RS codes for the P and Q sequences, respectively (see also Section 4.7). As the minimum distance $d_m = 3$ for both sequences, it will be possible to either detect exactly the location of 2 erroneous bytes within a codeword or directly correct one byte in error. Codewords including more than 2 bytes in error are detected with a certain probability. An alternative strategy will allow the correction of maximum 2 erroneous bytes when erasure-position information from the C2 corrector of the channel decoder is used. Depending on the correction strategy, the CD-ROM system approaches a bit error rate (BER) of $10^{-13} \ldots 10^{-15}$ when operating on *Mode 1* format. Further details about the third-layer error detection and correction for CD-ROM data can be found in [53,70,80,92].

The CDBD controller monitors all activities discussed until now. Usually, once the error correction has been finished, the resulting data is CRC-checked using the same 276-byte ECC field. Next, the host system initiates a DMA[9] process for transferring data from the block decoder. At this point, the task of the CDBD controller consists in a proper management of the data flow from the DMA buffer (FIFO) towards host.

A relatively new function of the block decoder is to supply the host with audio data at a higher bit rate than during normal audio read-out. This function has been incorporated in very high-speed CD-ROM drives (e.g. 32X) where there is no need anymore to spin the disc at 1X CLV for audio

---

[9]When using direct memory access (DMA), a disk-based storage device can transfer data straightforward to the computer memory, bypassing the central processing unit (CPU) of the host system [91,112].

playback. A dedicated input buffer collects the audio bytes delivered by the channel decoder and the controller transfers them, at another bit rate, to the host system. Nevertheless, because the audio symbols are not accompanied by a synchronization pattern, a special sync protocol[10] is needed between the channel decoder and CDBD.

As a final remark, we notice that modern block decoders are not only capable of handling CD-ROM *Mode 1* formats but also both *Mode 2* forms. This implies that other compact disc standards are also supported and data contained on various types of compact discs, like Video CD, Photo CD, CD-i, can also be read and forwarded to the host system.

6.4 CD-ROM volume and file structure

The information recorded on a CD-ROM disc should necessarily obey a collection of standardized rules, usually referred to as ISO 9660 standard [52]. This document specifies how computer data is placed on the disc. Any operating system which is supposed to read from a CD-ROM drive must be able to read this file structure.

ISO 9660 defines the arrangement of data in sectors within a volume and, viewed from a higher level, in files within directories. The related numbering and hierarchy, mode of identification, etc. are also described. The final result is that a CD-ROM disc looks for a computer user as a read-only hard-disk, including a bootable area. The storage capacity of a CD-ROM disc lies between 529 and 701 MB (see also Section 7.2).

Any group of 2048 user bytes originating from a CD-ROM data block and forwarded to the host computer is referred to as a sector. When viewed from the host side, the sectors are numbered in ascending order and mapped through addresses to a fixed table. The mapping process is valid throughout the whole disc and is called logical addressing. While retrieving data from disc, the CD-ROM block decoder converts the requested sector address into header information (i.e., physical address) and further into subcode timing.

---

[10]Usually, the beginning of an EFM frame (see Fig. 4.8) is used for audio synchronization between the channel and the block decoder. In addition, the subcode timing should also be provided to the CDBD controller.

Chapter 7

System Parameters and Drive Performance

The final performance of a CD-ROM drive is determined not only by the various technical choices adopted for the general architecture but, to a large extent, by the standardized parameters of the disc itself. In this context, this chapter will present an overview of several system parameters. The storage possibilities of a CD-ROM disc will also be discussed and some of the system choices adopted for the initial audio standard are listed. The chapter ends with a short description of the benchmarked drive specifications.

7.1 System parameters

A representative selection of standardized parameters [50,53,79,80] and calculated ones is listed in Table 7.1. If commonly used in calculations, the associated symbols are also indicated.

The compromises made more than 15 years ago to define a compact disc standard were of a various nature. The system limitations were closely related to physical and technological developments, as well as to the technical

Symbol	Parameter	Value	Unit
	Outer diameter of the disc	120 ± 0.3	mm
	Diameter of the center hole	$15^{+0.1}_{0}$	mm
	Disc thickness	$1.2^{+0.3}_{-0.1}$	mm
	Disc weight	$14 \ldots 33$	g
q	Track pitch	1.6 ± 0.1	μm
	Maximum spiral eccentricity [1]	± 70	μm
D_i	Starting diameter of the program area	$50^{0}_{-0.4}$	mm
D_o	Maximum diameter of the program area	116	mm
v_a	Linear velocity of the recorded data [2]	$1.2 \ldots 1.4$	m/s
λ	Wavelength of the laser light	780 ± 10	nm
NA	Numerical aperture of the objective lens	0.45 ± 0.01	–
n_{sub}	Refractive index of the transparent substrate	1.55 ± 0.1	–
	Disc reflectivity	$> 70\%$	–
L_{bit}	Length of the channel bit	$278 \ldots 324$	nm
	Pit depth	$60 \ldots 130$	nm
	Recording density	106	Mbit/cm^2
η_{rec}	Recording efficiency	28.4	%
	Maximum disc capacity	701	MB
S_{max}	Maximum playback time [3] at 1X CLV	$60 \ldots 80$	min
B_a	User bit rate at 1X CLV	153.6	kB/s
f_{ch}	Channel bit rate at 1X CLV	4.3218	Mbit/s
BLER	Block error rate before CIRC error correction	< 0.03	–
BER	Bit error rate after third-layer error correction	$< 10^{-12}$	–

[1] Relative to the disc center
[2] Also called scanning velocity (because the original CD-DA standard was defined for audio discs which were only played back at this speed) or reference velocity.
[3] Depending on disc tolerances

Table 7.1   CD-ROM system parameters.

possibilities for mass production. In fact, the situation has repeated itself quite recently, when the new standard for the digital versatile disc (DVD) was established.

7.2 Storage capacity

One of the main objectives to be achieved in a mass storage device is a high recording density. For an optical disk, in general, the number of data bits that can be stored on disc is given by [4,94]

$$N_b = \frac{\pi \left(D_o^2 - D_i^2 \right)}{4 \Phi_{spot}^2} \eta_{rec} \qquad (7.1)$$

where D_o and D_i are the maximum and respectively starting diameter of the program area, Φ_{spot} is the diameter of the laser spot, as defined in Section 2.3, and η_{rec} is the efficiency of the recording method. Clearly, the spot size needs being reduced in order to enhance the storage capacity.

The recording efficiency is a measure for the amount of bits delivered to the host system, as end user, with respect to the total amount of bits recorded on disc. As already indicated in Section 4.6, there are only 192 bits forwarded to the data path out of 588 channel bits. Further, the CD-ROM user has at his disposal only 2048 bytes out of a whole data block (see Section 6.2). The recording efficiency of a CD-ROM system becomes

$$\eta_{rec} = \frac{8 \times 2048}{98 \times 588} = 0.284 \qquad (7.2)$$

which means that any CD-ROM user bit, at the host interface level, originates from 3.52 bits recorded on the disc.

The capacity of a CD-ROM disc to store computer data is given by

$$C_{cdrom} = \frac{\pi \left(D_o^2 - D_i^2 \right)}{4 q L_{bit}} \eta_{rec} \cdot \frac{1}{8 \times 1024^2} \quad [MB] \qquad (7.3)$$

and current values between 529 and 701 MB can be obtained. The computer literature designates a CD-ROM disc as a 650-MB medium. Significantly to notice that, when the compact disc was standardized, the information capacity was not pushed to the optical limits in order to maintain a high reliability [97].

7.3 Design compromises

In designing a system, the desired specifications should carefully be weighted against each other. The final system performance depends on various factors, including physical tolerances, and they all need to be taken into account. The design process ends with a set of compromises which are thought to determine an optimal system.

Some of the trade-offs which led to the final compact disc standard are summarized below. Nevertheless, as most of the system parameters are interrelated, it is difficult to discuss a design compromise independent from the others, but a certain approach is still necessary.

Read-out of the disc in reflection. Basically, it is possible to have a transparent optical disc through which a laser beam is transmitted. The modulation of the laser beam in transmission is similar to the modulated reflection. However, the latter approach simplifies the player construction (electronics and optical components are situated at only one side of the disc), allows for shallower disc relief structure (pits/lands) which facilitates mass replication, and employs simpler methods for focus control [12]. The main advantage of the reflective read-out is the possibility to protect the information layer against dust, grease, and scratches.

Wavelength of the laser. The laser beam does not obey the approximations of geometrical optics [59,73] and is diffracted by the disc relief structure. As a consequence (see also Section 2.3), the spot will only have a diffraction-limited diameter [9] which linearly depends on the laser wavelength. A high recording density can be achieved by reducing this parameter but there is no really free choice with semiconductor lasers. In terms of mass production, a 780-nm GaAs device was already available at the time the CD-DA standard was proposed and it could offer the desired recording density.

Numerical aperture. As discussed above and indicated by Equation (2.1), the diffraction-limited laser spot size is proportional with $1/NA$. A higher numerical aperture is therefore needed to increase the recording density but also to improve the frequency response of the system [13]. On the other hand, a smaller numerical aperture leads to a larger focal depth, according to Equation (3.1), and to more relaxed tolerances for both disc thickness and skew [12,13]. When, in addition, the various optical aberrations are considered, a numerical aperture $NA = 0.45$ turns out to be a good system choice.

Track pitch. Due to diffraction, a certain amount of light falls outside the track being read out. A small fraction of the reflected light will therefore represent contributions from neighboring tracks, usually denoted as cross-talk. The track pitch plays an essential role while limiting this cross-talk but other factors, like quality of the scanning spot or the pit geometry, are equally important [12]. An ideal aberration-free spot yields very low cross-talk levels, of about -40 dB for a track pitch $q = 1.6\,\mu m$. A cross-talk level below -20 dB is generally accepted as requirement for audio and CD-ROM discs [97] but safe levels below -30 dB can normally be achieved, unless the spot suffers large aberrations [12].

Pit geometry. The optimum central aperture (CA) disc read-out takes place when the light reflected by a pit is approximately in antiphase with the light reflected by the surrounding area [12]. For a disc read in reflection through the transparent substrate ($\lambda = 780$ nm, $n_{sub} = 1.5$), this condition leads to a pit depth of $0.25\lambda/n_{sub} = 130$ nm. Typically, the pit depth is less than 130 nm to accommodate other system tolerances (e.g. variations of the laser wavelength) and eventually allow detection methods other than CA. The pit width, of about 0.6 μm, is determined partially by the optics and partially by the material properties of the disc [97]. The length of a pit depends both on optics and channel modulation, varying between $3L_{bit}$ and $11L_{bit}$ with L_{bit} being the length of the channel bit.

Channel modulation. The choice of the channel modulation affects the quality of the recovered signal, the frequency spectrum of the HF signal, and the length of a channel bit [44,93]. Modulation methods, like 8-to-8 or 8-to-24 were also analyzed, but EFM turned out to represent an optimal choice for the compact disc system.

Audio sampling frequency. Although not directly related to CD-ROM, this standardized parameter determines, in fact, the data throughput (see Sections 4.6, 5.4, and 6.2). For high-fidelity audio reproductions, a 20-kHz bandwidth is generally accepted as being sufficient but the sampling frequency of the original audio signal should be at least twice as high. The chosen value $F_s = 44.1$ kHz considers an additional reserve of 10 % for uniform reproduction of higher frequencies, including provision for post-DAC low-pass filtering. A second reason for choosing 44.1 kHz was the presence at that time on the market of digital audio recorders based on magnetic tape. These recorders were derived from the video equipment which used already 44.1 kHz for horizontal/vertical synchronization and it was thought that some compatibility towards newly-emerged CD system would be necessary [70].

Linear velocity of recorded data. The choice of this parameter was closely related to the frequency $F_s = 44.1$ kHz used to sample the analog audio signal as well as to the adopted modulation method (EFM). Based on the channel bit rate $f_{ch} = 588 \times 7350$ bit/s (see Section 4.6), the linear velocity of the recorded data is then given by $v_a = f_{ch} L_{bit}$.

Playing time of the disc. The total playback time of the disc is basically determined by the linear velocity v_a of the recorded data and by the total quantity of information stored along the disc spiral. The latter parameter depends on the track pitch, starting and maximum diameter of the program area, and on the length of the channel bit. However, it was agreed by Philips and Sony that a compact disc should be large enough to hold most of the long classical pieces of music [70]. The playing time of a standard Philips compact cassette was 1 hour and of the standard long-play (LP) record of the American company CBS was 20 to 30 minutes on one side. A compact disc had to hold at least 60 minutes of music but, on the other hand, a too large playing time was not really desired because the recorded disc would have costed too much (the royalties paid to the artists are proportional to the playing time).

7.4   Drive performance and benchmarking

The performance of a CD-ROM drive relies on three basic specifications: data transfer rate (also designated as bit rate, data throughput or simply data rate), access time, and CPU utilization.

The data throughput is determined by the speed at which data is transferred towards the host system (PC) and is measured at the host interface level. This specification is often indicated by the so-called X-factor and it has been the driving force behind the development of CD-ROM drives.

The access time gives an indication about the drive capability to find requested data and send it through the host interface as fast as possible.

The third important specification mentioned above concerns the time requested by a CD-ROM drive from the microprocessor of its host system to perform the necessary data access and transfer functions. The CPU utilization is mostly determined by the data path firmware (particularly by the buffer management) and by the CD-ROM driver[1] installed on the host computer.

Apart from the three main performance indicators listed above, the power consumption has also become an important issue. Higher disc rotational

---

[1] A device driver is a piece of software which gives the computer a handy set of control functions needed for various purposes [91]. Most device drivers are used to link the computer system (i.e., including software) to a given hardware device, e.g. a CD-ROM.

speeds combined with a very low access time require more current drawn from the power supply. In this respect, portable systems are those setting an upper limit for this performance specification.

Other drive specifications are more or less considered of secondary importance because they do not reflect directly the technical performance. Examples of such specifications are the mean time between failures (MTBF) or the weight of the drive and, clearly, they are not perceived by the user in a straightforward manner.

Nevertheless, the importance of some drive specifications may vary depending on a particular application. For instance, data throughput is relatively insignificant while playing back CD-ROMs with games or video information.

We shall only discuss here the data throughput and the access time, as being the only specifications which are solely determined by the drive itself and can be measured with a dedicated CD-ROM benchmark.

Although all building blocks (from disc to host computer) of the data channel have a contribution to the overall access performance, it is only the basic engine which determines the speed of seeking between two locations on disc. As for the sustained data throughput, it is again and only the basic engine which can improve this figure.

7.4.1 Data transfer rate

Usually, the data transfer rate of a disk drive is indicated separately for continuous read-out (sustained transfer) and as an absolute peak value [28, 29,30,75].

The transfer rate of a CD-ROM drive is strongly related to the disc standards [50,53,79,80]. Because data is uniformly recorded along a continuous disc spiral, data throughput will not change if the spiral is read out with a constant linear velocity (CLV). In turn, the disc rotational frequency will have to vary from the inner to the outer disc diameter according to Equation (4.6).

As already indicated at the end of Section 6.2, a CD-ROM disc rotating at 1X CLV delivers 153.6 kbytes per second and usually the numerical value $B_a = 150$ kB/s is adopted as reference. However, most current CD-ROM drives use either pure CAV turntable motor control or employ adaptive-speed techniques, like partial-CAV or zoned adaptive-speed. In these cases, the average sustained bit rate should be calculated by integration. When continuously playing back the whole disc, the average bit rate of the drive is given by [103]

$$\overline{B}_{av} = \frac{1}{S_{tot}} \int_0^{S_{tot}} B(S_x)\, \mathrm{d}S_x = \frac{B_a}{S_{tot}} \int_0^{S_{tot}} \mathcal{N}(S_x)\, \mathrm{d}S_x \qquad (7.4)$$

where S_{tot} is the total playback time of the disc (see Table 7.1) and $B(S_x)$, $\mathcal{N}(S_x)$ are the user bit rate and overspeed factor measured at the sub-code timing S_x, respectively. The maximum numerical value resulting from Equation (7.4) should be considered as design criterion when optimizing a CD-ROM drive for data throughput.

A benchmarking computer program, however, does not determine the average transfer rate while playing back the whole disc. In this case, the integration limits from (7.4) must be replaced by two values of the subcode timing S_x and S_y, between which the benchmark program reads out data.

7.4.2   Average access time

The term *average access time* describes the combination of average head positioning time and average disk rotation delay [28,29,30]. However, some CD-ROM drive manufacturers have fallen into the habit of using this term to describe average positioning time, or *seek time*. This usage does not adequately describe the time required for a disk drive to start responding to a host system request.

When a particular information is needed at the host level, the CD-ROM drive is set to access that information and retrieve it from the disc. The related access time [101] can be expressed as

$$T_{access} = T_{seek} + T_{mot} + T_{lat} + T_{ov} + T_r \qquad (7.5)$$

where the various contributions will be explained below. In general, the drive is characterized by an *average* access time which implies that all terms from Equation (7.5) should also be averaged. The significance of these terms can better be understood when plotting the overspeed profile during an outward-oriented seek, as schematically depicted in Fig. 7.1. A situation similar to the one from this figure occurs for an inward-oriented seek, but the overspeed profile displays a negative peak at the end of the seek action. Notice that a seek action should be regarded as a sledge displacement followed by track acquisition performed with the actuator.

Mechanical seek time. The numerical value of T_{seek} is determined by

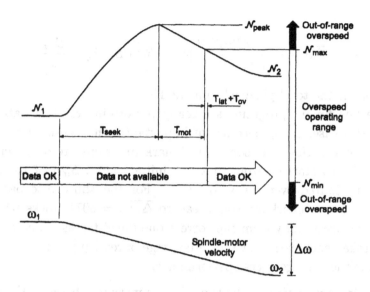

Figure 7.1   Overspeed profile and corresponding data availability, the
angular velocity of the turntable motor, and the compo-
nents of the access time for an outward-oriented seek.

the displacement of the optical pick-up unit[2] (OPU) from one location to
another along the disc radius. The time T_{seek} needed by the OPU to cross a
given number of tracks ΔN_{tr} can be obtained by solving the servo equations
in time domain. The number of tracks ΔN_{tr} is given by Equation (3.15).

The drive seek performance is usually represented by two indicators: the
third-stroke and the full-stroke seek times, respectively. It is important
to recall that a seek action, at the basic engine level, is directed to find
a subcode timing rather than a particular track. The average seek performance is therefore determined by the average number of tracks $\overline{\Delta N}_{tr}$
which are crossed while randomly seeking data. As an example [100],
$\overline{\Delta N}_{tr} = 6658$ tracks when the following disc parameters (see Table 7.1)
are considered: $D_i = 50$ mm, $D_o = 116$ mm, $q = 1.6$ μm, and $v_a = 1.3$ m/s.
Significantly to notice that $(D_o - D_i)/6 = 6875$ tracks for the same disc
parameters.

Also, because data is uniformly distributed along the disc spiral, the average seek length $\overline{\Delta S}$ expressed in time units between two randomly chosen
values S_x and S_y of the subcode timing is given by [100]

---

[2]See Sections 3.1 and 3.6.

$$\overline{\Delta S} = \frac{1}{S_{tot}^2} \int_0^{S_{tot}} \int_0^{S_{tot}} |S_x - S_y| \, \mathrm{d}S_x \mathrm{d}S_y = \frac{S_{tot}}{3} \qquad (7.6)$$

where S_{tot} is the total playback time at 1X.

Some benchmark programs are really performing random data access[3] and, in this case, $\overline{\Delta N}_{tr}$ becomes of importance when designing the radial seek controller. However, some benchmark programs are only performing seeks of $S_{tot}/3$ starting from different initial values S_{init} of the subcode timing distributed over the whole disc. For the above disc parameters, this type of benchmarking would lead to $\overline{\Delta N}_{tr} = 6071$ tracks which does not differ substantially from the correct one (6658 tracks). These so-called third-stroke benchmarks have the advantage of reducing the time needed to determine the average access performance.

Turntable motor time. The time T_{mot} is measured between the moment at which the target track is found through mechanical displacement and the moment at which data read-out becomes possible (see Fig. 7.1).

As a component of the access time, T_{mot} does not exist in other disk storage devices like hard-disk or floppy-disk drives. In CD-ROM systems, however, the turntable motor time is intrinsically related to the standardized CLV distribution of data along the disc spiral which gives rise to an over- or underspeed while accessing data. For example, the overspeed N_{peak} measured a the end of an outward-oriented seek may exceed the maximum overspeed N_{max} which can be handled by the channel decoder (and data path). The consequence is a delay T_{mot} which is introduced until the turntable motor brakes to a safe angular frequency.

This delay depends on the turntable motor itself but, very important, on the type of the motor control being used. The minimization of T_{mot} can be assured by properly choosing the speed profile and, if necessary, strongly accelerating/braking the turntable motor during (and shortly after) mechanical displacement.

The calculation of the average turntable motor time for a given strategy of the motor control depends primarily on the overspeed profile which sets the values of N_1 and respectively N_2 before and after any (random) seek. However, while continuously accessing data, the average value of T_{mot} may become zero because the motor speed neither arrives at a steady state nor

---

[3]The benchmark program generates random logical addresses (see Section 6.4) which are converted by the data path into physical addresses and further into subcode timing.

reaches its target velocity [105].

Latency time. The latency T_{lat} is defined as the time needed for the desired information (situated along the disc spiral) to arrive under the laser spot while the disc is rotating. The latency is measured from the moment at which data read-out becomes possible after finding the target track through mechanical displacement.

On average, the target subcode is located 180° from the position of the laser spot. When compared to other disk-based storage devices [66], the latency in CD-ROM systems is not necessarily a constant value because the rotational frequency of the disc depends on the type of the motor control being used. For a given subcode-dependent overspeed profile $N(S_x)$, the average latency becomes

$$\overline{T}_{lat} = \frac{\pi}{v_a S_{tot}} \int_0^{S_{tot}} \frac{1}{\mathcal{N}(S_x)} \sqrt{\frac{D_i^2}{4} + \frac{v_a q S_x}{\pi}} \, dS_x \qquad (7.7)$$

which turns into

$$\overline{T}_{lat} \bigg|_{clv} \approx \frac{\pi(D_o + D_i)}{4 \mathcal{N} v_a} \qquad (7.8)$$

and

$$\overline{T}_{lat} \bigg|_{cav} = \frac{1}{2 f_{rot}} = \frac{\pi D_i}{2 \mathcal{N}_i v_a} \qquad (7.9)$$

when the spiral is read out at constant linear velocity (CLV) and constant angular velocity (CAV), respectively. The parameter N_i designates the overspeed at the starting diameter of the program area while the disc is rotating in CAV mode.

Overhead time. As the target subcode is found along the disc spiral, it will still take a time T_{ov} until data is sent through the host interface.

Part of this additional time, amounting for about 1-5 ms, is due to firmware overhead and can be reduced, for example, by increasing the microcontroller clock and by correctly structuring the firmware code. Other delays are introduced by the hardware processing which needs 98 demodulated frames (13.3 ms at 1X) to obtain one value of the subcode timing and 111 demodulated frames[4] (15.1 ms at 1X) to deliver one CIRC-corrected user frame. Further, data path needs 98 user frames to deliver one 2048-byte sector. These delays, which are fixed at 1X by the compact disc standards [50,53,79,80], decrease proportional with the overspeed factor when

---

[4]The longest symbol delay during CIRC decoding.

the disc is played back at higher scanning velocities.

Retry time.   Due to various factors, including tolerances of the disc parameters, disc unbalance, etc., the target track may not be reached at the end of a solely sledge-actuator displacement (seek). In such cases, a retry accounting for T_r is needed. Reducing the average retry time to a negligible value represents one of the most challenging design criteria in current high-speed CD-ROM drives.

Bibliography

[1] Alonso, M. and Finn, E.J. *Physics*. Addison-Wesley Publishing Company, 1992.

[2] Arai, T., Okamoto, H., Nishimura, K., Kobayashi, M, and Takeuchi, T. "High Capability Error Correction LSI for CD Player and CD-ROM". *IEEE Transactions on Consumer Electronics*, Vol. CE-30, No. 3, pp. 353-359, Febr. 1984.

[3] Baggen, C.P.M.J. and Nijhof, J.E. "A Compact Disc IC Concept Enhancing Playability". *IEEE International Conference on Consumer Electronics. Digest of Technical Papers*, June, 1985, pp. 48-49.

[4] Bartolini, R.A. "Optical Recording: High-Density Information Storage and Retrieval". *Proceedings of the IEEE*, Vol. 70(6), pp. 589-597, June 1982.

[5] Bartolini, R.A., Bell, A.E., Flory, R.E., Lurie, M., and Spong, F.W. "Optical Disk Systems Emerge". *IEEE Spectrum*, August, 1978, pp. 20-28.

[6] Bergmans, J.W.M. *Digital Baseband Transmission and Recording*. Kluwer Academic Publishers, 1996.

[7] Best, R.E. *Phase-Locked Loops. Design, Simulation, & Applications. Third Edition*. McGraw-Hill, Inc., 1997.

[8] Bhattacharya, P. *Semiconductor Optoelectronic Devices*. Prentice-Hall International, Inc., 1994.

[9] Born, M. and Wolf, E. *Principles of Optics. Electromagnetic Theory of Propagation, Interference and Diffraction of Light. Sixth Edition*. Pergamon Press, Oxford, Reprinted 1993.

[10] Botez, D., Spong, F.W., and Ettenberg, M. "High-Power Constricted Double-Heterojunction AlGaAs Diode Lasers for Optical Recording". *Applied Physics Letters*, Vol. 36, No. 1, 1980, pp. 4-6.

[11] Bouwhuis, G. and Braat, J.J.M. "Video Disk Player Optics". *Applied Optics*, Vol. 17, pp. 1993-2000, 1978.

[12] Bouwhuis, G., Braat, J., Huijser, A., Pasman, J., Rosmalen, G. van, and Schouhamer Immink, K. *Principles of Optical Disc Systems.* Adam Hilger Ltd., Bristol and Boston, 1985.

[13] Braat, J. "Optics of Recording and Read-Out in Optical Disk Systems". *Japanese Journal of Applied Physics*, Vol. 28 (Suppl. 28-3), pp. 103-108, 1989.

[14] Braat, J. and Bouwhuis, G. "Position Sensing in Video Disk Readout". *Applied Optics*, Vol. 17(13), pp. 2013-2021, July 1978.

[15] Braat, J.J.M. and Bouwhuis, G. "Optical Video Disks with Undulating Tracks". *Applied Optics*, Vol. 17, pp. 2022-2028, 1978.

[16] Bricot, C., Lehureau, J.C., and Puech, C. "Optical Readout of Videodisc". *IEEE Transactions on Consumer Electronics*, pp. 304-308, November 1976.

[17] Bulthuis, K., Carasso, M.G., Heemskerk, J.P.J., Kivits, P.J., Kleuters, W.J., and Zalm, P. "Ten billion bits on a disk". *IEEE spectrum*, August, 1979, pp. 26-33.

[18] "Byte - 20th Aniversary Report". *Byte*, September 1995, pp. 54-114.

[19] Candy, J.C. and Temes, G.C. "Oversampling methods for A/D and D/A Conversion". *Oversampling Delta-Sigma Data Converters. Theory, Design, and Simulation*, IEEE Press, 1992.

[20] Carasso, M.G., Peek, J.B.H., and Sinjou, J.P. "The Compact Disc Digital Audio System". *Philips Technical Review*, Vol. 40, No. 6, pp. 151-155, 1982.

[21] Carlson, A.B. *Comunication Systems. An introduction to Signals and Noise in Electrical Communication. Third Edition.* McGraw-Hill, New York, 1986.

[22] *CD-ROM Professional. The Magazine for CD-ROM Producers & Users.* 1995-1999, Online, Inc., Wilton, Connecticut, USA.

[23] Chait, Y., Park, M.S., and Steinbuch, M. "Design and Implementation of a QFT Controller for a Compact Disc Player". *Journal of Systems Engineering*, Vol. 4, pp. 107-117, 1994.

[24] Clark Jr., G.C. and Cain, J.B. *Error-Correction Coding for Digital Communications.* Plenum Press, New York, 1981.

[25] Coops, P. "Mass Production Methods for Computer-Generated Holograms for CD Optical Pickups". *Philips Journal of Research*, Vol. 44, No. 5, pp. 481-500, 1990.

[26] Couch II, L.W. *Digital and Analog Communication Systems. Fifth Edition.* Prentice-Hall International. Inc., Upper Saddle River, NJ 07458, 1997.

[27] Data Books: Hitachi Semiconductor. *Hitachi ASSP for Audio Applications,* 1993; Philips Semiconductors. *Semiconductors for Radio and Audio Systems,* 1997; Rohm Co., Ltd. *Data Book. Optical Disc/Magnetic Disk ICs,* 1997-1998; Sanyo. *Audio-Use MOS Integrated Circuits,* 1993; Sony. *Optical Disk,* 1994-1995; Toshiba. *Audio Digital ICs,* 1992.

[28] *Disk/Trend Report. Optical Disk Drives.* 1993-1997, DISK/TREND, Inc., Mountain View, California, USA.

[29] *Disk/Trend Report. Rigid Disk Drives.* 1995-1997, DISK/TREND, Inc., Mountain View, California, USA.

[30] *Disk/Trend Report. Removable Data Storage.* 1995-1997, DISK/TREND, Inc., Mountain View, California, USA.

[31] Doebelin, E.O. *Measurement Systems. Application and Design. Fourth Edition.* McGraw-Hill, Inc., 1990.

[32] Dutton, K., Thompson, S., and Barraclough, B. *The Art of Control Engineering.* Addison Wesley Longman, 1997.

[33] *EMedia Professional. The Magazine for Electronic Media Producers & Users.* 1995-1997, Online, Inc., Wilton, Connecticut, USA.

[34] Enden, A.W.M. van den, Verhoeckx, N.A.M. *Discrete-Time Signal Processing. An Introduction.* Prentice Hall International (UK) Ltd., 1989.

[35] Fukunaga, N., Yamamoto, M., Kubo, M., and Okabayashi, N. "Si-OEIC (OPIC) for Optical Pickup". *IEEE Transactions on Consumer Electronics*, Vol. 43, No. 2, pp. 157-164, May 1997.

[36] Gardner, F.M. *Phaselock Techniques*. John Wiley & Sons, Inc., New York, 1966.

[37] Gates, B. *The Road Ahead*. Penguin Books USA, Inc., 1995.

[38] Gill, P.E., Murray, W., and Wright, M.H. *Practical Optimization*. Academic Press, London, 1981.

[39] Goedhart, D. and Dijkmans, E.C. "Low-Pass Filter and Output Circuits for Compact Disc". *IEEE International Conference on Consumer Electronics. Digest of Technical Papers*, June, 1985, pp. 46-47.

[40] Goodman, J.W. *Introduction to Fourier Optics*. McGraw-Hill, Inc., 1968.

[41] Goto, K., Higuchi, Y., Kume, M., Ohgoshi, S., Yamada, A., and Suzuki, A. "Laser Pickup for CD Player". *IEEE International Conference on Consumer Electronics. Digest of Technical Papers*, June, 1985, pp. 44-45.

[42] Gray, P.R. and Meyer, R.G. *Analysis and Design of Analog Integrated Circuit Design. Second Edition.* John Wiley & Sons, New York, 1984.

[43] Hartmann, M., Jacobs, B.A.J., and Braat, J.J.M. "Erasable magneto-optical recording". *Philips Technical Review*, Vol. 42, No. 2, pp. 37-47, 1985.

[44] Heemskerk, J.P.J. and Schouhamer Immink, K.A. "Compact Disc: System Aspects and Modulation". *Philips Technical Review*, Vol. 40, No. 6, pp. 157-164, 1982.

[45] Hoeve, H., Timmermans, J., and Vries, L.B. "Error Correction and Concealment in the Compact Disc System". *Philips Technical Review*, Vol. 40, No. 6, pp. 166-172, 1982.

[46] Hopkins, H.H. "Diffraction Theory of Laser Read-Out Systems for Optical Video Disc". *Journal of the Optical Society of America*, Vol. 69(1), pp. 4-24, January 1979.

[47] Horowitz, P. and Hill, W. *The Art of Electronics. Second Edition.* Cambridge University Press, New York, 1990.

[48] Howe, D.G. "CD Error Characterization; Differences between CD-ROM and Writable CD". *Proceedings of Topical Meeting on Optical Data Storage*, May 16-18, 1994, Data Point, California.

[49] *HÜTTE. Die Grundlagen der Ingenieurwissenschaften. 29 Auflage.* Springer-Verlag, Berlin, 1989.

[50] International Electrotechnical Commission. *Compact Disc Digital Audio System.* IEC 908. First Edition, 1987.

[51] Isailović, J. *Videodisc and Optical Memory Systems.* Prentice Hall, Inc., 1985.

[52] *ISO 9660. Information Processing - Volume and File Structure of CD-ROM for Information Interchange. First Edition,* 1988.

[53] *ISO/IEC 10149. Information Technology - Data Interchange on Read-Only 120 mm Optical Data Disks (CD-ROM). Second Edition,* 1995.

[54] Jurgen, R.K. *Digital Consumer Electronics Handbook.* McGraw-Hill, 1997.

[55] JVC, Mathsushita Electronics, Philips Electronics N.V., and Sony Corp. *Video CD Specifications. Version 2.0,* 1994.

[56] Kahlman, J.A.H.M. and Baggen, C.P.M.J. *Apparatus for Furnishing Data Signals from an Optically Readable Record Carrier at the Same Frequency as the Associated Subcode Signals.* U.S. Patent no. 4.631.714, December 1986.

[57] Kenney, G.C., Lou, D.Y.K., McFarlane, R., Chan, A.Y., Nadan, J.S., Kohler, T.R., Wagner, J.G., and Zernike, F. "An Optical Disk Replaces 25 Mag Tapes". *IEEE Spectrum*, February, 1979, pp. 33-38.

[58] Klapper, J. and Frankle, J.T. *Phase-Locked and Frequency-Feedback Systems.* Academic Press, Inc., 1972.

[59] Klein, M.V. and Furtak, T.E. *Optics. Second Edition.* John Wiley & Sons, Inc., 1986.

[60] Kreyszig, E. *Advanced Engineering Mathematics. Seventh Edition.* John Wiley & Sons, Inc., 1993.

[61] Kuo, B.C. *Digital Control Systems. Fifth Edition.* CBS Publishing Asia, Ltd., Hong Kong, 1987.

[62] Kuo, B.C. *Automatic Control Systems. Sixth Edition.* Prentice-Hall International, Inc., New Jersey, 1991.

[63] Kurata, Y., Yamaoka, H., Ishikawa, T., Coops, P., Duivestijn, A., and Zoeten, P. de. "CD Optical Pickup Using a Computer-Generated Holographic Optical Element". *Proceedings of the Society of Photo-Optical Instrumentation Engineers. Optical Storage and Scanning Technology,* Vol. 1139, pp. 161-168, 1989.

[64] Marchant, A.B. *Optical Recording. A Technical Overview.* Addison-Wesley Publishing Company, 1990.

[65] Maréchal, A. and Françon, M. *Diffraction, Structures des Images.* Masson et Cie, Paris, 1970.

[66] Mee, C.D. and Daniel, E.D. *Magnetic Recording. Volume II: Computer Data Storage.* McGraw-Hill, Inc., 1988.

[67] Mendenhall, W. and Sincich, T. *Statistics for Engineering and the Sciences.* Prentice Hall, Inc., 1995.

[68] Meyr, H. and Ascheid, G. *Synchronization in Digital Communications. Volume 1. Phase-, Frequency-Locked Loops, and Amplitude Control.* John Wiley & Sons, Inc., 1990.

[69] Moon, J.H., Lee, M.N., and Chung, M.J. "Track-Following Control for Optical Disk Drives Using an Iterative Learning Scheme". *IEEE Transactions on Consumer Electronics,* Vol. 42, No. 2, pp. 192 – 198, May 1996.

[70] Nakajima, H. and Ogawa, H. *Compact Disc Technology.* Ohmsha, Ltd., 1992.

[71] Naus, P.J.A. *Bitstream Digital-to-Analogue Conversion for Digital Audio.* Master of Philosophy Thesis, University College of Swansea Wales, United Kingdom, 1991.

[72] Naus, P.J.A., Dijkmans, E.C., Stikvoort, E.F., McKnight, A.J., Holland, D.J., and Bradinal, W. "A CMOS Stereo 16-bit D/A Converter

for Digital Audio". *IEEE Journal on Solid-State Circuits*, Vol. SC-22, pp. 390-395, 1987.

[73] O'Shea, D.C. *Elements of Modern Optical Design.* John Wiley & Sons, Inc., 1985.

[74] Papoulis, A. *Signal Analysis.* McGraw-Hill, Inc., 1977.

[75] *PC Magazine.* 1993-1997, Ziff-Davis, Inc., Boulder, Colorado, USA.

[76] Pedrotti, F.L., Pedrotti, L.S. *Introduction to Optics.* Prentice-Hall, Inc., 1993.

[77] Peek, J.B.H. "Communications Aspects of the Compact Disc Digital Audio System". *IEEE Communications Magazine*, February, 1985, Vol. 23, No. 2, pp. 7-15.

[78] Peterson, W.W. and Weldon Jr., E.J. *Error-Correcting Codes. Second Edition.* The MIT Press. Massachusetts Institute of Technology, 1972.

[79] Philips Electronics N.V. and Sony Corp. *Compact-Disc Digital Audio. System Description*, 1981.

[80] Philips Electronics N.V. and Sony Corp. *Compact-Disc Read-Only Memory. System Description*, 1984.

[81] Philips Electronics N.V. and Sony Corp. *System Description CD-ROM XA*, 1991.

[82] Plassche, R. van de. *Integrated Analog-to-Digital and Digital-to-Analog Converters.* Kluwer Academic Publishers, 1994.

[83] Pohlmann, K.C. *Principles of Digital Audio. Third Edition.* McGraw-Hill, Inc., 1995.

[84] Poor, A. "DVD and CD-ROMs. 21st Century Storage". *PC Magazine*, January 21, 1997, p. 164-181.

[85] Pozo, L.F. *Glossary of CD-ROM Technology.* Issued by the Special Interest Group for CD-ROM Applications and Technology, Reston, VA 22303-1324, USA, 1996.

[86] Proakis, J.G. and Manolakis, D.G. *Digital Signal Processing. Principles, Algorithms, and Applications. Second Edition.* Macmillan Publishing Company, 1992.

[87] Prosise, J. "Everyone's Guide to Computer Acronyms, Part I and II". *PC Magazine*, January 9, 1996, p. 259-266 and January 23, 1996, p. 211-217.

[88] Rankers, A.M. *Machine Dynamics in Mechatronic Systems. An Engineering Approach*. PhD Thesis, Twente University, The Netherlands, 1997.

[89] Redheffer, R. and Port, D. *Differential Equations. Theory and Applications*. Jones and Barlett Publishers, Boston, 1991.

[90] Rohde, U.L. *Digital PLL Frequency Synthesizers. Theory and Design*. Prentice-Hall, Inc., Englewood Cliffs, New Jersey, 1983.

[91] Rosch, W.L. *Hardware Bible, Premier Edition*. SAMS Publishing, Indianapolis, 1997.

[92] Sako, Y. and Suzuki, T. "CD-ROM System". *Proceedings of Topical Meeting on Optical Data Storage*, October 15-17, 1985, Washington, D.C.

[93] Schouhamer Immink, K.A. *Coding Techniques for Digital Recorders*. Prentice Hall International (UK) Ltd., 1991.

[94] Schouhamer Immink, K.A. "The Digital Versatile Disc (DVD): System Requirements and Channel Coding". *SMPTE Journal*, pp. 483-489, August 1996.

[95] Schouhamer Immink, K.A. "EFM Coding: Squeezing the Last Bits". *IEEE Transactions on Consumer Electronics*, Vol. 43, No. 3, pp. 491 − 495, August 1997.

[96] Schouhamer Immink, K.A. and Gross, U. "Optimization of Low-Frequency Properties of Eight-to-Fourteen Modulation". The Radio and Electronic Engineer, Vol. 53, No. 2, pp. 63-66, February 1983.

[97] Shannon, R.R., Wyant, J.C. *Applied Optics and Optical Engineering. Volume IX*. Academic Press, 1983.

[98] Shepherd, W., Hulley, L.N., and Liang, D.T.W. *Power Electronics and Motor Control. Second Edition*. Cambridge University Press, 1995.

[99] Stan, S.G. *Study on (2,7) PPM Bit Detection*. Master of Electronic Engineering Thesis, Eindhoven International Institute, Eindhoven University of Technology, The Netherlands, 1992.

[100] Stan, S.G. "Twin-Actuators for Ultra-Fast Access in CD-ROM Systems". *IEEE Transactions on Consumer Electronics*, Vol. 42, No. 4, pp. 1073-1084, Nov. 1996.

[101] Stan, S.G., Akkermans, A.H.M., Steinbuch, M., and Norg, M.L. "A System Analysis for High- and Very High-Speed CD-ROM Drives". *IFAC International Workshop on Motion Control*, Grenoble, France, September 21-23, 1998.

[102] Stan, S.G. and Bakx, J.L. *Device for Reading and/or Recording Information on a Disc-Shaped Information Carrier*. U.S. Patent pending, April 1995.

[103] Stan, S.G. and Bakx, J.L. "Adaptive-Speed Algorithms for CD-ROM Systems". *IEEE Transactions on Consumer Electronics*, Vol. 42, No. 1, pp. 43-51, Febr. 1996.

[104] Stan, S.G., Kempen, H. van, Leenknegt, G., and Akkermans, A.H.M. "Look-Ahead Seek Correction in High-Performance CD-ROM Drives". *IEEE Transactions on Consumer Electronics*, Vol. 44, No. 1, pp. 178-186, Feb. 1998.

[105] Stan, S.G., Kempen, H. van, Lin, C.C.S., Yen, M.S.M., and Wang, W.W. "High-Performance Adaptive-Speed/CAV CD-ROM Drive". *IEEE Transactions on Consumer Electronics*, Vol. 43, No. 4, pp. 1034-1044, Nov. 1997.

[106] Stan, S.G., Yen, M.S.M., Lin, C.C.S., Wang, W.W., and Cheng, K.C. "Adaptive-Speed/CAV Algorithm in a CD-ROM Drive to Accomplish High Data Transfer Rate and Low Power Consumption". *Asia-Pacific Data Storage Conference*, July 16-18, 1997, Taoyuan, Taiwan.

[107] Steinbuch, M and Norg, M. "Industrial Feedback". *Fourth European Control Conference ECC97*, July 1-4, 1997, Brussels, Belgium.

[108] Steinbuch, M., Schootstra, G., and Bosgra, O.H. "Robust Control of a Compact Disc Player". *Proceedings of the 31st Conference on Decision and Control*, December, 1992, Tucson, Arizona.

[109] Sze, S.M. *Semiconductor Devices. Physics and Technology.* John Wiley & Sons, Inc., 1985.

[110] Takeuchi, T., Saitoh, T., Komatsu, S., and Nakamura, T. "CD-ROM Image Retrieval System". *Proceedings of SPIE. Optical Mass Data Storage II*, August 18-22, 1986, San Diego, California.

[111] "The Windows Evolution". *Windows Magazine*, December 1995, pp. 228-240.

[112] Tischer, M. *PC Intern. System Programming.* Abacus, Grand Rapids, MI, USA, 1994.

[113] Tsunoda Y., Sawano, S., Nakamura, H., Saito, K., Tsukada, T., and Takeda, Y. "Semiconductor Laser Pickup for optical Video Disk Player". *IEEE Transactions on Consumer Electronics*, Vol. CE-23, No. 4, 1977, pp. 479-493.

[114] Veldhuis, R. and Breeuwer, M. *An Introduction to Source Coding.* Prentice Hall International (UK) Ltd., 1993.

[115] Velzel, C.H.F. "Laser Beam Reading of Video Records". *Applied Optics*, Vol. 17, No.13, pp. 2029-2036, 1978.

[116] Watkinson, J. *The Art of Data Recording.* Focal Press, Butterworth-Heinemann Ltd., Oxford, 1994.

[117] Wicker, S.B. and Bhargava, V.K. *Reed-Solomon Codes and Their Applications.* IEEE Press, 1994.

[118] Williams, A.B. and Taylor, F.J. *Electronic Filter Design Handbook.* McGraw-Hill, Inc., New York, 1995.

[119] Williams, E.W. *The CD-ROM and Optical Disc Recording Systems.* Oxford University Press, Inc., New York, 1994.

[120] Wilson, J. and Hawkes, J.F.B. *Optoelectronics: An Introduction.* Prentice-Hall International, Inc., 1983.

[121] Xiaoping, H., Sigh, G., Weerasooriya, S., and Low, T.S. "Transfer Function Modeling of the Focus, Radial and Sledge Actuator in a CD-ROM Drive". *Asia-Pacific Data Storage Conference*, July 16-18, 1997, Taoyuan, Taiwan.

[122] Yamauchi, H., Miyamoto, H., Sakamoto, T. Watanabe, T., Tsuda, H., and Yamamura, R. "A 24X-Speed CIRC Decoder for a CD-DSP/CD-ROM Decoder LSI". *IEEE Transactions on Consumer Electronics*, Vol. 43, No. 3, pp. 483 – 490, August 1997.

[123] Young, M. *Optics and Lasers. An Engineering Physics Approach.* Springer-Verlag, Heidelberg, 1977.

[124] Zadeh, L.A. "Optimality and Nonscalar-valued Performance Criteria". *IEEE Transactions on Automatic Control*, Vol. AC-8, p. 1, 1963.

Index